Simple Solution Essays

by PATRICK KENJI TAKAHASHI

authorHOUSE®

AuthorHouse™
1663 Liberty Drive
Bloomington, IN 47403
www.authorhouse.com
Phone: 1-800-839-8640

First published by AuthorHouse 9/7/2010

ISBN: 978-1-4520-7176-3 (e)
ISBN: 978-1-4520-7175-6 (sc)

Library of Congress Control Number: 2010912386

Printed in the United States of America

This book is printed on acid-free paper.

Dedication

This book is dedicated to my wife, Pearl. She passed away on 19July2009. After returning from the hospital that day, I decided to write an article for **The Huffington Post** on my experience over the past five weeks when she was a patient there. You will find this piece in the essays as "Gratitude, Not Grief."[1]

The Pearl Foundation was created as a tribute to her life. She so loved a special yellow tree that a search ensued to identify that specie. It turned out to be **Tabebuia Donnelli-Smithii** or **Cybistax DT**, as shown on the back cover. These trees will be planted at sites throughout Hawaii in remembrance of her. Thus, the cover color selection as yellow.

---

1   http://www.huffingtonpost.com/patrick-takahashi/gratitudenot-grief_b_241390.html

SIMPLE SOLUTION ESSAYS

Table of Contents

Introduction

This is Book 3 of my series on *SIMPLE SOLUTIONS*.[2] Book 1 focused on Planet Earth and Book 2 on the rest of my life, or Humanity. Almost all these essays were published in *The Huffington Post*, rated by *Technorati* as the #1 blog site.[3] A few did not fit the requirements of *HuffPo*, so they come from my daily blog.[4]

You will note that I have chosen to use footnotes instead of a gigantic reference at the end. Further, shown are link sites to enter the information age. It's a bit cumbersome to re-copy the whole address, but that is no different from a standard reference. In time, this book will be placed into the world wide web to simplify the process. [Hint: the essays will be serialized in by daily blog.]

I also chose to just follow these essays in chronological order, so that the import of that moment could unfold over time. I provide a short history of why each essay was written. Each ends with a review of comments (the number refers to responses). You can, of course, go to a particular *HuffPo* site footnoted to read the details. They ranged from zero comments to more than a hundred.

Clearly, there is a virtual, but real, connection between my blog and *HuffPo*, for on 27February2010, I was asked by them to submit an article on the coming tsunami to Hawaii. That 8.8 moment magnitude earthquake had just struck Chile, and there was some anxiousness about gigantic waves. I said, sure, be glad to be of service…however, I happened to be in Amsterdam. No problem, for CNN had a live feed from Waikiki Beach and Hilo Harbor, and I was more up on what was happening in Hawaii than most living this experience, for many were up in the hills awaiting the possible doom. You can read that posting, but I mention all this because my daily blog averages just under a hundred hits/day. That day, 3,356 visited my blog.

I would look forward to any feedback. Try:

ptakahas@hotmail.com

Aloha.

---

2   http://simplesolutionsbook1.com and http://simplesolutionsbook2.com

3   http://technorati.com/blogs/top100

4   http://planetearthandhumanity.blogspot.com

My very first Huffington Post article was published on 29May2008 at a time when U.S. Senators Hillary Clinton and Barack Obama were competing for the Democratic Party nomination for president. Very simply, I did not think that Hillary, with her strong ties to the establishment, and John McCain, a Republican, had any chance of setting the foundation for universal peace. Barack was new and wanted change. What better gift to Humanity than to set the stage for World Peace.

Well, Barack, We have a Problem...[5]

...and only you will have the power to provide the solution. Why you? The United States is the most powerful nation, ever. Today, and for the next generation, no other country will be anywhere close to our military and economic dominance. You will have a once in a millennium opportunity to accomplish something monumentally extraordinary, while ameliorating the global economic mega-depression that some say will soon loom from the combined hammer of Peak Oil and Global Warming.

You worry that you have a more important task at hand, which is to become POTUS (President of the U.S.)? Yes, continue that effort, but you will become the 44th POTUS.

So, on to the more important challenge, creating that ultimate legacy for the benefit of Planet Earth and Humanity. As an aside, let me say that during the past year I published two books: **SIMPLE SOLUTIONS for Planet Earth** (http://SimpleSolutionsBook1.com) and **SIMPLE SOLUTIONS for Humanity** (http://SimpleSolutionsBook2.com). I mention this because those publications provide both the spur and solution for you, our Nation and the World.

Our society seems to have a fatal flaw: we can't expeditiously act to prevent potential cataclysms like Peak Oil and Global Warming (let's call this PO/GW). Yes, of course, we will eventually prevail, but, by my reckoning, only after decades of agony and turmoil. There has to be a better way.

Miracle of miracles! It turns out, ironically, that PO/GW is just the catalyst you need to empower you to take those remarkable steps. There are innumerable great things you can attempt to accomplish as the POTUS. What about something so supreme as ending wars... forever?

Let us speculate that early in your presidency your close advisors tell you, Mr. President, we have a problem. We have reached Peak Oil, and the $100+/barrel oil we have will soon zoom to over $200/barrel. It seems, also, that we can't shift to coal or other fossil fuels to manufacture a liquid fuel because, yikes, Global Warming is indeed real. It's kind of too late, and you can ascribe culpability to Bush the 43rd or Congress or oil executives, but, to be

---

5   http://www.huffingtonpost.com/patrick-takahashi/well-barack-we-have-a-pro_b_104201.html

honest, the masses--you and me and others--are mostly to blame. Public will is totally lacking on energy policy.

Understanding that, you and you alone can immediately orchestrate a global Manhattan/Apollo effort to minimize the coming pain. We need to spend a trillion dollars over the next few years to stimulate industry to help us remediate the almost certain crisis. This is only a fraction of what we will squander on the wars in Afghanistan and Iraq, so the numbers are tolerable to take on this gallant mission.

So where do you find this kind of money? Ah, the Defense budget. You go to your very first G8 Nations Summit, by your declared emergency to be held at United Nations headquarters in New York City, and pronounce a Gorbachev-like bombshell: our country will reduce military spending by 10% this year, and will continue to do so for the next eight years. This scenario is described on page 65 of my **Book 2**.[6] You say, we want every country to do the same, for this 10% solution is our best response to Peak Oil and Global Warming. At this stage, keep quiet about the "ending wars forever" part, as then, no one will take you too seriously.

China's knee-jerk reaction might well be, what, cut defense spending? We haven't had a chance yet to attain your level of capability. But, on afterthought, they will realize that they will only need to decrease their spending by $6 billion in Year One while the U.S. takes a $60 billion hit. Ten percent of the worldwide $1.2 trillion/year for war means that at least $120 billion/year will suddenly become available in the first year to overcome PO/GW. This accessible sum will drop to just under $100 billion in Year Two...and down to a little more than $50 billion by Year Eight. But, by Year 12, the world defense budget will have been reduced to $34 billion (a little more than 5% our current DOD budget), and almost a trillion dollars would have been allocated to overcome PO/GW. That will be the Year 2020, providing perfect vision for Planet Earth and Humanity.

This grand sum will go to the United Nations to administrate and dispense. Yes, I've worked with the UN, and it is about the worst organization to do anything, but there is no choice, it is the only international entity of any credibility available.

The so-called military-industrial complex will shift their effort to mass-producing more efficient wind energy conversion systems, developing the hydrogen jetliner, in time converting the carbon dioxide in the atmosphere to methanol and otherwise insuring for the development of the sustainable infrastructure and necessary clean energy. Military personnel can initially be maintained to handle the few thousand terrorists, take on environmental tasks and the like. By 2020 there should one U.S. security force not more than 5% our current size.

There is every reason to believe that your 10% solution can continue forever to the point where there will be close to no military expenditures and, therefore, no chance for a major world war. Countries like Costa Rica, Iceland, Mauritius and Panama already have no defense budget. Are they threatened? Nope! Japan and China have had a chance to expand their

---

6   http://simplesolutionsbook2.com

economy because of a limited war account. The U.S spends almost $2000/person for defense while that of China is $45/citizen.

Triumphing over PO/GW will mean applying our tax dollars in a constructive manner. Better jobs will be created, greater support will be applied to education and the world economy will truly be boosted.

So, future President Obama, you will have a problem, but could the above scenario be an effective solution? You are the exact right person to save Planet Earth and Humanity. Simply go to my blog at http://PlanetEarthAndHumanity.blogspot.com, where some details are presented.

Ah, what a great country. Where else can a former public McKinley High School student provide advice to a Punahou graduate?

Comments (15): _The respondents were almost all supportive. One even called the 10% solution a fantastic plan. There was one statement about my naiveté, but my readers jumped into it by criticizing this negative attitude._

That first HuffPo was, of course, totally ignored, so on to my second of 2June08, which was about as prosaic as anything I write can be. However, it must have struck a nerve, for there were more than a hundred comments. The price of oil was skyrocketing, and there was nothing we could do to stop it.

Why Is There No National Energy Policy?[7]

There is no national energy policy because there is no public will for one. Sure, blame Congress or President Bush or the oil companies, but we are the problem. We, meaning you, me and others. Part of the reason for our relative insouciance is that life remains okay.

For example, European gasoline prices can be almost triple ours. France is just below $10/ gallon, Norway just above and Germany is already up to $11.50. As would be expected, protests are growing. Soon, there could be major demonstrations, if not uncontrolled rioting. You can, of course, move to Venezuela. *Simple Solutions for Planet Earth* listed gasoline at 12 cents / gallon last year. Today, gasoline still costs 12 cents / gallon there. But do you really want to relocate to that part of the world just for gasoline?

By the way, ten years ago, the price of crude oil was $11.91/barrel, less than one-tenth the cost today. Has something monumental occurred? Ah... yes, although my response should have been more hysterical.

When you come down to it, we have our priorities all wrong, and do nothing about righting this nonsense. For example, each NASA space shot is said to cost about a billion dollars. This is more than the annual Department of Energy renewable energy budget.

Each B-2 Stealth bomber sells for one billion dollars... no, make that $2 billion, including all development costs. A B-2 weighs 2.3 million troy ounces, which, if made of pure gold, would then have a value of, yes, about $2 billion. Of course, it won't fly, but want more?

The U.S. Navy will outlay $160 million/year to man each Nimitz class aircraft carrier, and when the George Bush is christened (no joke, but named for POTUS #41, H.W.) this year, there will be ten of them. The President George W. Bush solar energy budget request is about half what it will cost to operate one nuclear carrier! And we have ten of them... with no major enemy today, and none clearly on the horizon for the next generation or more.

The American public readily accepts this absurdity. Where are the priorities? What can you do about this? You can start by reading those two *Simple Solutions* books.

The world needs to spend not billions, but trillions, of dollars over the next few years to minimize the crunch of Peak Oil and Global Warming (PO/GW). Nobel Laureate Joseph

---

7   http://www.huffingtonpost.com/patrick-takahashi/why-is-there-no-national_b_104507.html

Stiglitz has a new book entitled, **The Three Trillion Dollar War**, reporting on the true cost of the Middle East war. The PO/GW 10% solution recommended to POTUS #44 in my May 29th post[8] only provided the global federal government investment. Most of the actual outlay will need to come from industry. But, without the force of law or appropriate spur from government, we have seen that corporations are loath to move into unexplored investment areas, as, for example, sustainable resources, or the remediation of global warming.

So with crude oil settling in the range of $126/barrel (which is exactly $3/gallon), are we now, finally, soon to get a national energy policy? No. Two presidential candidates (John McCain and Hillary Clinton) have actually proposed eliminating the 18.4¢/gallon (24.4¢/gal for diesel) tax during the three summer months to win some votes. Barack Obama said it was a gimmick and Thomas Friedman of **The New York Times** entitled his editorial:[9] "Dumb as We Wanna Be." Friedman said, in reference to McCain and Clinton, "the unifying idea is so ridiculous, so unworthy of the people aspiring to lead our nation, it takes your breath away."

What's the reality? First of all, if this tax is dropped for the summer, the consumer would be saving 5% -- 5% -- on their gas bill. This should be a non-issue! Gasoline increases this amount each week, sometimes, if not more.

How much money are we talking about? Actually, something in the range of $10 billion. Wow, that's a lot. The U.S. Department of Energy spent less than this amount for renewable energy R&D, cumulative, over the past decade.

Well, Exxon, during this past quarter, made a profit of $10.9 billion, which Wall Street found disappointing. Chevron's profits only amounted to $5.2 billion, the second best quarter in their 132 year history. (How long has oil been around, anyway? Well Chevron first struck oil in California in 1876. Drake drilled for his Pennsylvania petroleum in 1859.) Remember, these profits are only for January, February and March of 2008.

Cutting the gas tax is going in the wrong direction. We need to add a dollar/gallon and apply this revenue to developing sustainable options. That would provide about $150 billion each year, which would be just the complement to the 10% solution suggested in my first article. National energy policy? Why bother? A dollar/gallon investment tax on gasoline, the 10% solution and a 10 cents / pound carbon dioxide tax are all we need. Carbon tax? Stay tuned.

Comments (103): The response was mostly supportive. Interestingly enough, as of the Summer of 2010, we still don't have a national energy policy, and there will be none for some time to come.

---

8   http://www.huffingtonpost.com/patrick-takahashi/well-barack-we-have-a-pro_b_104201.html

9   http://www.nytimes.com/2008/04/30/opinion/30friedman.html?_r=2&oref=slogin

*On 8June08, it was time to look at Peak Oil and Global Warming. Petroleum was already up to $138.54/barrel, and unemployment had jumped, setting the stage for the economic collapse in the Fall. Incidentally, Charles Krauthammer is a Pulitzer Prize columnist for the **Washington Post**. He is also a central conservative, meaning, he rails on Democrats, but not as vitriolically as, say, Ann Coulter. Of course, the Hokkaido G8 Summit went nowhere and no action was taken, except to plead with the OPEC countries to produce more oil. Amazing!*

Preaching To The Choir[10]

Is global warming a liberal and academic conspiracy? Yes, says Charles Krauthammer (http://nationalreview.com) in his 30May08 editorial. *The Wall Street Journal* on June 7 reported that the Copenhagaen Consensus Center, featuring a blue-ribbon panel of economists with five Nobel Laureates, picked climate change mitigation as #30, and last, of thirty world priorities. The Journal did not mention that Bjorn Lomborg, the founder of the Center, is author of The *Skeptical Environmentalist* and noted critic of global warming. We've all seen liberal equivalents. How, then, to get both sides to interact and work together?

First, the bad news, for the past few days have been ominous and depressing with regard Peak Oil and Global Warming:

1. At the national level, our unemployment rate saw the biggest monthly jump in 22 years, the price of petroleum experienced the highest one day increase of $11/barrel to an all time high of $138.54/barrel, the Dow Jones plummeted by nearly 400 points, and the U.S. Senate killed the climate change bill.

2. The International Energy Agency reported that a sum of $45 trillion (not billion, trillion) will be required to combat global warming.[11] Then, of course, there was that Scandinavian announcement of the insignificance of climate mitigation.

So what do they all mean? First, there are credible individuals and organizations sincerely at odds. Why did the Democrat-controlled Senate pass on the climate? Republicans are reluctant to take the problem too seriously and Democrats thought that the legislation was lukewarm at best. That is decision-making? Anyway, cap and trade is a compromise. A real carbon tax has to be the answer. All in all, not terrific, but not bad, either.

Second, the combined hammer of Peak Oil and Global Warming is finally beginning to affect the global economy, as stock markets are beginning to dive and joblessness is rising. Protests about energy and food prices are sprouting. Worse, it is appearing that our national policy promoting ethanol from corn is exacerbating the situation.

---

10   http://www.huffingtonpost.com/patrick-takahashi/preaching-to-the-choir_b_105943.html

11   http://www.reuters.com/article/idUSL0348323120080603

Third, the reasonably cautious International Energy Agency suggesting $45 trillion to do something about climate change is sobering, as the total annual world defense budget is only $1.2 trillion. That's bad enough, but decision-makers are not even close to agreeing on any societal response. Well, on to Japan in July, for the G8 nations will supposedly focus on the environment. Don't hold your breath, though, for the previous meeting only led to a nuclear agreement and some consensus about climate change becoming an issue around mid-century. The world energy ministers, kicking off the Hokkaido Summit, reacted to the jump in oil prices by tepidly pleading with OPEC to increase production. President Chakib Khelil immediately responded with a sure we'll consider your input...when we next meet in September.

Okay, how, then, to bring all callings to operate as a unified choir? There is a fatal flaw in our society: we can't make strategic decisions until it is too late. I've tried with inspiration (see my posting on "Well, Barack, we have a problem...) and fear (***SIMPLE SOLUTIONS for Planet Earth***, Chapter 5, "The Venus Syndrome"). Maybe what we need is good, old fashioned, diplomacy. While it might be too late by then, POTUS #44 Obama will host the G8 Summit in that upcoming fateful year, 2012, and, perhaps, the world might then be ready for his ultimate challenge.

Comments (0): Yes, there were no comments.

On June 9 I thought I would begin linking global climate change with Peak Oil, as the first thought of decision-makers would be to replace oil with coal. However, coal results in double the carbon dioxide in the atmosphere, so a two-by-four approach appeared to be necessary to scare politicians away from this fossil option. I needed to do something extreme, because our Congress had just killed the global warming mitigation bill, and, remember, George W. Bush was still our President. Thus came the worst case scenario:

The Venus Syndrome (Part One)[12]

If you've been keeping score, just during the past week:

1. Oil experienced the biggest one day price increase, ever, and ended up at a record high of $138.54/barrel. G8 oil ministers kicking off the next summit in Japan begged OPEC to increase production.

2. Our monthly unemployment rate saw the largest jump in 22 years.

3. The International Energy Agency reported that a sum of $45 trillion (not billion, but trillion) was needed to combat global warming.

4. The U.S. Senate killed the global warming mitigation bill.

5. The Copenhagen Consensus Center, involving a blue-ribbon panel of economists, including five Nobel Laureates, listed global warming as #30, and last, of 30 world priorities.

6. The **Huffington Post** expanded to include the Green Section.

Let's see now, on the one hand, the feared dual hammer of Peak Oil and Global Warming is already showing evidence of affecting our lives. Yet, our decision-makers are continuing to do nothing.

Actually, as would be expected, things are a bit more complicated than that. For example, the congressional legislation was passed on to next year because Democrats felt that the current bill was not good enough, and that Scandinavian organization was founded by the author of **The Skeptical Environmentalist**, who has become famous for panning climate change.

Thus, from all indications, global warming has become almost believable... yet not compelling enough for decision-makers to take any truly preventative steps. Most would say that the problem is the ole USA, with China and India comfortable in being excluded from the Kyoto Protocol, for why should they be concerned when we--the largest carbon dioxide polluter--do not wish to be cooperative?

---

12 http://www.huffingtonpost.com/patrick-takahashi/the-venus-syndrome-part-o_b_106120.html

Clearly, President George W. Bush and VP Dick Cheney are a prime reason why our nation is so recalcitrant, although industry barons generally also feel that coal, our dominant energy resource, surely can't be as bad as the liberal environmentalists make it out to be. I don't think that the need to maximize profits is a major reason why corporate boards are generally lukewarm on global warming and opposed to something like the carbon tax. The primary reason is that this group, while some do show concern, likes to think that there is no cause for alarm at this time.

Charles Krauthammer, an avowed agnostic on the Greenhouse Effect, is typical of this currently ruling group. His **National Review Online** article of May 30, 2008 paints the problem as a liberal conspiracy. He could well be right, but I'm of the opinion that something must be done now or it could become too late.[13]

I've tried in sundry ways. I participated in forty years of sustainable energy and environmental R&D, and remain quasi-active in an effort on biological hydrogen production. But I needed to try other mechanisms, so last year I applied the "Einstein writing to FDR to build the atomic bomb"[14] model by enlisting a few media colleagues in an attempt to get President Bush to take this problem more seriously, and, hopefully, surprise the world by taking a leadership role in the then upcoming G8 summit scheduled for Germany by proposing a post-Kyoto strategy for climate change.

First, we couldn't find an Einstein, so we thought a message from top corporate officials would suffice. Well, we started with General Motors and got nowhere. What could you expect from a company that has a global vice president like Bob Lutz on staff? He has been quoted to say that climate warming is a crock of sh*t, or something similar. Scratch this strategy.

Well, pure logic was not working, so I tried something even more offbeat--FEAR! I wrote a book (SIMPLE **SOLUTIONS for Planet Earth**), with Chapter 5 on "The Venus Syndrome." What are some strategies?

What about sea level increase as a scare tactic? If all the ice melts, the oceans will rise by about 250 feet. Wow, that should strike terror into decision-makers. Yeah, but this could take milennia to actually happen. Yes, it could be up to a meter by the end of this century, but that's an extreme high, and the more probable tenth of an inch every few years can safely be ignored. We will feel sorry for those poor souls living on threatened atolls, but will not get too excited about something so infinitesimal at our coastline. Anyway, our authorities will more probably build walls around coastal cities rather than cut out fossil fuels, something known as the Iron Lung Syndrome—which is to treat the symptom, but avoid the root of the problem.

Ah, but it turns out that there is another greenhouse gas, methane, that shows promise for becoming the doomsday gas. Methane is the simplest hydrocarbon and is most of natural gas.

---

13 http://climateprogress.org/2008/05/30/krauthammers-strange-denier-talk-points-part-1-newtons-laws-were-overthrown/

14 http://www.dannen.com/ae-fdr.html

Cows burp and flatulate this gas, while biomass generally also produces some in the decay process.

We have all been schooled to plant trees. When you do this, what happens? They eventually die, and, yes, the carbon dioxide is returned into the atmosphere, but, worse, the decomposition process also produces some methane. There was even a recent German study that seemed to hint that growing trees produced more methane than we thought, so the greening of lands might well actually hurt our environment. Stay tuned!

So what's the big deal? Well, one molecule of methane is from 20 to 60 times worse than one molecule of carbon dioxide in causing global warming. This miniscule amount of methane in our atmosphere already has half the potency of carbon dioxide in warming our globe.

Let's take the case of our planetary neighbor. Venus is mostly carbon dioxide at a surface temperature of almost 900 degrees Fahrenheit. Could you imagine a scenario where sufficient methane contaminates our atmosphere to really cause trouble? Methane tends to oxidize into carbon dioxide over time. Ergo, the potential for an atmosphere and surface temperature on Earth like Venus.

We tend to ignore the ocean in our scientific analyses. Did you know that there is more combined mass in the bacteria, viruses and archaea in the ocean than in all the larger life forms (fish, trees, you) in the ocean and on land? You probably never even ever heard of archaea.[15] The scary thing is that marine microorganisms at the surface expire, drop to the bottom of the ocean, and in an absence of oxygen, are converted into methane and other compounds, which, because of the pressure and temperature at depth, generally become trapped in ice as marine methane hydrates. It is said that there might be twice the energy in this methane at the seabed than all the known coal, oil and natural gas. Let me repeat: TWICE AS MUCH ENERGY IN METHANE IN METASTABLE EQUILIBRIUM AT THE BOTTOM OF THE OCEAN THAN ALL THE KNOWN COAL, OIL AND NATURAL GAS DEPOSITS, WHICH ARE RATHER SAFELY RESTING DEEP UNDERGROUND.[16]

Over our geologic history, every few tens of million years, our planet naturally heats up. This is accompanied by heightened carbon dioxide and methane levels, or more probably, these gases caused the temperature rise...just like today. Some scientists have speculated that the primary cause might well have been a rather sudden release of marine methane hydrates into the atmosphere. Tomorrow, Part Two of THE VENUS SYNDROME will appear, providing a tale from the future on what might happen to Planet Earth if methane goes haywire.

Comments (10): There was a fun series of exchanges discussing the merit of fear as a tactic and the potential reality of marine methane hydrates.

---

15   http://www.ucmp.berkeley.edu/archaea/archaea.html

16   http://marine.usgs.gov/fact-sheets/gas-hydrates/title.html

The Venus Syndrome (Part Two) appeared the following day (10June08), and provided a worse case scenario.

The Venus Syndrome (Part Two)[17]

Yesterday, Part One reported on the science and politics of global warming. Today, we enter the realm of the unknown, the future. Remember, now, there could be twice as much energy content in methane trapped at the bottom of the ocean than all the known coal, oil and natural gas deposits.

What happens to gas in melting ice? What happens to gas and ice in water? Well, the latter floats to the surface and they both enter the atmosphere. What if a combination of circumstances jiggles all the marine methane hydrates to the surface? You can add Arctic tundra to this mix, as it, too, is full of methane, and already melting. What if there is a worst case scenario of coincidences? Such is my tale from the future--**THE VENUS SYNDROME**:

October 21, 2012, was a bad day for Planet Earth, in fact, the worst day ever for homo sapiens since Mt. Tubo erupted 71,000 years ago, dropping our human population to 15,000 or so. The temperature had finally reached 110°F in Washington, D.C., breaking the record set last year by 2°F. And so late in the year, too! But the temperature itself was not the problem; it was more the accumulation of the past three years being the hottest for the entire globe. Global climate warming was not only happening, but getting worse.

Hurricane Valerie was storming to the District. Only three Category 5 hurricanes have hit the U.S. Mainland (Camille in 1969, Andrew in 1992 and Igor in 2010) in the fifty years prior to this, but already three have struck this year (Debby in May and Tony last month). However, Hurricane William, churning in the Gulf of Mexico, and seeking New Orleans, has the lowest minimum pressure ever of 878 mb (millibars, with the previous low being an unnamed 1935 hurricane at 892 mb), with sustained winds of up to 200 miles per hour. The Army Corps of Engineers warned that funds were not made available to shore up the dikes beyond Category 3 after Katrina, so only the worst is anticipated.

Of particular concern to scientists, though, is that the thermohaline circulation might have stopped today. These currents kept certain coastal regions cool and others warm. They also helped maintain the marine methane hydrate (MMH) deposits below critical temperature. Expected now is a steep increase in ocean temperatures in many of those shelf areas heavy in MMH deposits.

---

17 http://www.huffingtonpost.com/patrick-takahashi/the-venus-syndrome-part-t_b_106325.html

Both the north terrestrial and magnetic North Poles have been over water for the past couple of months, with the seafloor 13,000 feet (4000 meters) below. Normally, only half the ice melts. As of today, there is no ice in the Arctic around the poles.

The Intergovernmental Panel on Climate Change (IPCC) met last month in an emergency session to declare that the permafrost of the U.S., Canada and Russia is, indeed, melting, and the 10.4°F (5.8°C) temperature rise by 2100 will now occur by 2025. The International Arctic Research Center (IARC) of the University of Alaska at Fairbanks reported that the northern ecosystem holds one fourth to one third the world's soil carbon, and much of it is carbon dioxide and methane trapped in the ice. Over the past 200 years, the atmospheric concentration of methane has tripled. Now, this ice is thawing with the deeper solid turning into slush, and, the expectation is that the effect of methane on global climate warming will in a few years exceed that of carbon dioxide. A 2°C increase in permafrost temperature amplifies methane flux by 120%, and there is a real danger that taliks, or thawed permafrost, are disintegrating, to further accelerate the effect.

The IARC also will soon release the results of a joint Russia-U.S. cruise investigating methane plumes found in the East-Siberian Sea. The speculation is that methane gas hydrates are melting at the bottom of the sea. As there is one million times more natural gas (methane) in these permafrost reservoirs than the normal methane annually released from the northern ecosystem, the implications are frightening.

Ten years ago (Alaskan Science Forum, January 3, 2001, Article #1523 by Ned Rozell), it was reported that 50% of Russia and Canada, 80% of Alaska, 20% of China and all of Antarctica is underlain by permafrost. In northern Siberia this permafrost is 1,600 meters (5,250 feet, or a mile) thick. A decade ago, this permafrost had warmed to within one degree Celsius of thawing in Alaska. Major melting was expected by 2040. It is occurring now.

The global system then even further positively reacts by melting the permafrost, contributing more carbon dioxide and a lot more methane into the atmosphere, which speeds up the process. Even an idiot would be able to determine that this cascading effect is not a smart thing to foster.

All this heat finally resulted in portions of the ocean going hyper-critical with respect to temperature, and the danger point for runaway water vapor was finally exceeded on this October day in 2012. Water vapor, as prevalent as it is, has not in the past been considered as a danger, mostly because there is so much of it and there is nothing we can do about it anyway. Suddenly, this is a major issue. But that in itself was not the trigger, for it should have taken many millennia to get to anything like the Venus Syndrome.

The problem was that this critical condition occurred over portions of the warmer Pacific, which sat over huge deposits of marine methane hydrates (MMH). It was purely coincidental, but in the Ring of Fire, at a spot off Peru, a major subsea earthquake rated at 8.9, triggered a massive underwater volcanic eruption, which served as a fuse to destabilize the hydrates, beginning the release of copious amounts of methane into the atmosphere. A major tsunami is expected to hit the Pacific Rim, but that has now become a minor irritant. A sizable MMH deposit off Guatemala, possibly catalyzed by a related earthquake, also went metastable. Unexpectedly, there was a huge reservoir of methane gas below the clathrates that just came to the surface.

Thus, in a matter of 24 hours, Planet Earth had its hottest day in modern history, a sudden influx of methane from the ocean, and an atmosphere where methane superseded carbon dioxide as the primary agent for global warming. We're not quite sure how this excess water vapor was affecting the process, but it can only be bad. The Great Ocean Conveyor Belt stopping was particularly ominous, for the ocean surrounding these MMH beds would further warm.

It had to take the Global Warming equivalent of a Perfect Storm to catalyze an expedited Venus Syndrome: portions of the ocean surface at critical positive feedback temperature; cessation of the cooling currents thus warming the marine methane hydrate (MMH) deposits; a major subsea earthquake combined with a cataclysmic undersea volcanic eruption; resultant tsunami which slightly lowered the sea level over the MMH deposits near the coastlines; and, most importantly, crossing over of the dynamic equilibrium pressure-temperature condition allowing marine methane to explode to the surface. Yes, and that troublesome water vapor influx.

Scientists at the International Marine Methane Hydrates Research Institute at the University of Hawaii calculated that one teraton, a million times a million, of methane will be released into the atmosphere over the next year or two. The early Eocene, 55 million years ago, experienced a similar event, resulting in a temperature rise of about 16 degrees Fahrenheit across the now populated regions of the world.

An emergency session of the United Nations, panicked by these developments, was called for this 21st day of October with a wary eye on Valerie. Members of the IPCC were summoned to participate. Computer models over the past week were refined for the most probable case scenario, and the terminal report was presented to the General Assembly at 10AM that fateful Monday. New York City, badly enough, was already at 102°F that morning, with a dangerous hurricane approaching, but that all palled to the devastating news.

The chairman of the IPCC reported that we had reached and tripped over the tipping point. THE VENUS SYNDROME had begun, and, now, could probably not be stopped. Humanity at large, and most of life, would cease to exist within a century, providing a short period to develop solutions for survival, the only rational one being to leave Planet Earth, although emergency efforts are being planned to release air pollution particulates and sulfate aerosols into the atmosphere. Fortunately enough, the Search for Extraterrestrial Intelligence (see Simple Solutions for Humanity in box on the right) project, headed by a non government organization, had recently detected signals from an apparently advanced civilization in the Orion constellation. The data is being interpreted and....

A long time ago, according to Greek mythology, there was a very beautiful princess of Troy by the name of Cassandra. Her younger brother Paris was the character who kidnapped Helen, the wife of the King of Sparta, and brought her to Troy. Apollo, son of Zeus, fell in love with Cassandra and gave her the power to know the future if only she were to marry him. She was given that power, but refused to marry him, so Apollo put a curse on her predictive capabilities, and doomed her to despair, for while her powers remained, no one would believe her.

Sometimes I feel that I have Cassandra's curse. Who knows? Look for *The Venus Syndrome*: the Novel, soon to come.

But not to fear, for while humanity might be sacrificed, Planet Earth will survive, find a way to re-gain balance, unlike Venus, and dedicate another 5 billion or so years to re-initiate life, about the time it took to produce us. But what are the odds?

Oh, the simple solution? Check out those **SIMPLE SOLUTION** books. Or maybe it's hopeless and there is now more substance to that Mayan prophecy about the year 2012.

Comments (12): *With a couple of wishful thinkers and giver uppers, the discussion was supportive and underscored the gravity of marine methane hydrate upheaval and tundra melting. Stay tuned for* **The Venus Syndrome**: *The Novel, perhaps in 2011. Yet do I have the Cassandra Curse?*[18]

---

18 http://www.gold-eagle.com/editorials_05/swagell041006.html

This must have been a productive period for me, as also on 10June08, the following biofuel posting was published comparing ethanol and methanol. The U.S. and world had jumped headfirst into ethanol, and I thought that was stupid and foolish.

Ethanol Vs. Methanol[19]

Ethanol and biodiesel are dead, long live methanol! Methanol is the simplest alcohol, with one carbon atom; ethanol has two. Thus, given biomass, it should be cheaper to produce methanol than ethanol. Surely enough, in a comprehensive assessment Stone & Webster performed for the U.S. Department of Energy two decades ago, with the Hawaii Natural Energy Institute as an associate, this fact was confirmed.

However, methanol has a few flaws. First, if drunk, you can go blind. But, who drinks gasoline? Second, there was a time when methanol was used as the feedstock to produce MTBE as a gasoline additive. MTBE is carcinogenic. Methanol is not, just don't drink it. Third, methanol can dissolve certain plastics and embrittle some metals. So change the plastic and metals to avoid this problem.

Methanol has only half the energy content per gallon of gasoline. Ethanol is two-thirds the intensity of gasoline. However, a fuel cell powered vehicle is at least twice the efficiency of an internal combustion engine, so the tank storage problem would be solved with a direct methanol fuel cell (DMFC). The DMFC for portable electronics is said to soon replace batteries, so the technology is real. Methanol is the only biofuel capable of being directly fed to a fuel cell. Ethanol and gasoline need to first be passed through an expensive reformer.

Plus, and this is difficult to accept, but true: one gallon of methanol has more hydrogen than one gallon of liquid hydrogen. Thus, the infrastructure is already largely in place for a methanol economy. George Olah in his book, ***Beyond Oil and Gas: The Methanol Economy,***[20] provides all the science and speculation you need.

So why is our country and rest of world enamored over ethanol and biodiesel? In two words, the Farm Lobby. They came up with a politically brilliant scheme to use corn as an answer to imported oil. By so doing, the price of farm commodities recently doubled and more. Farmers are ecstatic! The poor around the world are suffering.

Global food riots occurred, so the Farm Lobby thought, oh, no problem, we'll now, more and more, begin to convert the cellulose into ethanol, for, after all, those tax incentives are already in place. Well, if you have biomass and want a biofuel, you either hydrolyze and ferment it to produce ethanol, or gasify and catalyze it to make methanol. But the current

---

19   http://www.huffingtonpost.com/patrick-takahashi/ethanol-vs-methanol_b_106380.html

20   http://www.amazon.com/Beyond-Oil-Gas-Methanol-Economy/dp/3527312757

mentality is stuck in an ethanol mode. Before farmers and their partners build fermented ethanol from biomass factories, they need to totally re-think the long term and just change the congressional language to say: ethanol, biodiesel and other renewable biofuels. Methanol does not even need to be mentioned. Otherwise, they will be creating a second herd of white elephants.

With all this logic, won't methanol soon displace ethanol? No. Why? The Farm Lobby is so dominant that they will continue to insure for the continued use of ethanol for another decade because those facilities are already built, and they don't want them to suddenly become obsolete. Okay, fair enough, let those plants profitably phase out. But don't compound the problem by adding that second elephant herd.

I might add that there has been a sudden surge of interest in biofuels from algae. Certainly, as algae can be from two to ten times more efficient in converting sunlight into biomass than any terrestrial crop; grown in the ocean where there is no irrigation problem (and Peak Freshwater looms on the horizon); if fed the cold water effluent from the ocean thermal energy conversion process there will not be a need for fertilizers (deep ocean effluents are high in just the right nutrients--farm fertilizers are manufactured from fossil fuels); and with genetic engineering, who knows where this option can go--this has been my dream for a third of a century. However, the eventual costs are unknown. Yes, do the R&D, but don't expect a magic solution within a decade. Biomethanol is real and immediately available for commercial prospecting.

As no one I know is commercially jumping unto the methanol bandwagon, I will tomorrow publish a hypothetical letter to colleagues to inspire some enterprise. The strategies, then, become available to the readers of the **Huffington Post**. Also, too, perhaps some partnerships can be stimulated to come up to a better solution than ethanol and biodiesel. Let's do more than share ideas. Let's take action!

<u>Comments (11)</u>: *There were some detractors with the same old story about how dangerous this alcohol can be. These tend to be ethanol promoters, so they naturally have a bone to pick about a competitor. I noticed this same sort of feedback on several of my global warming articles, and after some tracking, learned that they work for organizations like the Heartland Foundation, funded by fossil fuel money. Professionals follow these publications and weigh in with their disinformation.*

On 12June08 I posted a follow-up article to see if there was any interest from the investing public.

What Is The Best Biofuel?[21]

This is Part Two regarding biofuels. Part One compared ethanol with methanol. This earlier analysis summarily dismissed biodiesel as far too inefficient and miniscule to even be considered for attention.

I've spent most of my professional life in academia. However, my early years were devoted to biomass engineering in industry, and, over time, I also gained broad experience from three years as a Special Assistant in the U.S. Senate, helping start several companies and serving on the board of Hawaii Biotech during its important transition period. For about a third of century I have in various capacities been involved in a wide range of processes to convert biomass into a liquid fuel. I've chaired a range of bio-energy conferences and for the specific field of methanol from biomass, assisted in securing funds in the neighborhood of $25 million to conduct experiments and participate in economic assessments of this option.

You can refer to the biomass section of Chapter 2 of **SIMPLE SOLUTIONS for Planet Earth** for the details, but by all common, economic and scientific sense, the simplest of alcohols, methanol, should be the sustainable fuel of choice over ethanol. I was thrilled when Nobel Laureate George Olah published **Beyond Oil and Gas: The Methanol Economy**. All the science and future speculation you wish to know can be found in this book about this alternative.

Thus, after a lifetime of research and pontification, I recently called my own bluff and drafted a letter to several colleagues to interest them in establishing a biomass to methanol holding company, to be called BioMethanol International. If a partnership is formed, we would seek the advice and active partnership of individuals and organizations throughout the world. If anyone more enterprising can be influenced to start his own company using the following strategy, great, as the whole point of this article is to get this field going.

In the following "hypothetical" letter, A, could well be an experienced venture capitalist operating out of Manhattan, NYC, and B might be Dr. Methanol, himself, perhaps a former professor at MIT who once ran the National Renewable Energy Laboratory (NREL) biomass gasification program. C and D are university faculty members. E could be a consultant with the Department of Energy and F a high level official of NREL. In principle, we've been talking about taking this step for over a long, long time, so while these alphabets represent real people who are all aware of the contents, nothing official has happened, yet. So, here is that open letter to the readership of the *HuffPo Green*:

Dear A, B, C, D, E and F:

---

21   http://www.huffingtonpost.com/patrick-takahashi/what-is-the-best-biofuel_b_106574.html

Ethanol and biodiesel are slowly sinking as biofuel options, but the Farm Lobby will ensure that, if we decide to form a methanol producing enterprise, we will have a few years to refine the effort. As far as I can determine, there is no group obviously on the horizon spearheading a similar venture. E mentioned to me that the bio-energy planning session held in Honolulu in May did not even consider this pathway. While F indicated to me a few months ago that NREL was hoping to start discussions about expanding the national biomass program, it appears that USDOE headquarters will be sticking to their current policy of purposefully excluding methanol. All this is mostly good, for the lack of competition is an almost necessary requirement for us to proceed.

Shockingly, I haven't seen even one overview paper treating the topic: okay, ethanol and biodiesel are dead, so, what else is there? British Petroleum seems to be stuck on fermentation for higher carbon biofuels, a very slow and inefficient process, and Shell is dabbling in cellulosic ethanol (if you have fiber, it is easier and cheaper to produce methanol), hydrogen and marine algae for diesel. If there are other teams around the world at our stage or beyond already proceeding, who are they? Maybe we can link with one of them.

The timing is ideal, then, to quickly form a holding company to pursue this methanol from biomass initiative. I would like to suggest the following ten step strategy:

1. Select a company name. We can always change it to suit our needs, so, for now, it shall be BioMethanol, International. If anyone has a better idea, let us know.

2. A, you are the only financial guy on our team, so you can be President. I'm writing this, so I'll make myself chairman of the board. B, do you want to be chairman of the Scientific Advisory Committee? C and D, you're teaching, but you can serve on B's committee if you wish. If you can suggest anything more relevant, we'd like to hear from you. E, as you're still consulting for the USDOE, we need to keep you on an informal advisory capacity. Anyone know others who might serve on either the company board or the scientific committee?

3. This should be simple boiler-plate formality for you, so, A, can you send us your version of the articles of incorporation and any financial details, including a strategy for angel financing.

4. We need to begin the process of adjusting congressional language to qualify methanol for the existing and future tax incentives. I'll look into this.

5. C and D, if you wish to participate, can you review the state of knowledge? Gasification systems, catalysts, whatever. You're all busy, so something simple would be satisfactory. If any developmental pathway looks especially promising, let's co-op the technology by asking them to join our partnership, but this has to be after we look credible enough. We should also, later, add a process for converting marine macroalgae to methanol. The moisture content could be a problem. Japan is ahead of us in this area, so I'll inquire.

6. We need to already interact with Barack Obama's people to insure that in the transition they will better appreciate the need to find another pathway to liquid fuels, and methanol,

more specifically, should be in their vocabulary. Discussions must be held with the staff of our congressional delegates. We can discuss how this initiative might unfold, but, E, can you figure out who our government contact should be? As of this moment, there is no individual allowed to even think methanol. This person has to be important in 2009, if she exists.

7. There might well be an oil company or equivalent interested in bankrolling this effort. We should explore around, for they have tens of billions available for their future. Under any circumstances, we will someday soon need a major player involved.

8. What our holding company will do will be to sign up the most promising technologies, meld them into an operational system and commercialize the process. Of course, considerable research will still be necessary to fill in the gaps and otherwise enhance the above, so C and D can handle those plans as necessary.

9. I'm completely open to anything, but my sense is that we will create the best possible biomass to methanol concept, including at least a pilot plant operation, and hope some larger company buys us out in five years. A more successful version of the Maui biogasifier project, one that produces real methanol, needs to become operational within five years for a sum of, say, $25 million to $50 million.

10. The toughest part of our challenge will be to secure funding. Well, actually, the more difficult requirement will be to convince anyone of importance that methanol makes more sense than ethanol. We need to discuss these matters at our next stage of activity. While most of this can be accomplished electronically, it would be ideal if a critical mass of us can meet for a day within a few months, either in Honolulu or Denver.

I have attached two articles delving into methanol, and, in particular, the direct methanol fuel cell, for this liquid is the only biofuel capable of being fed to a fuel cell without reformation. I think the DMFC for vehicles will be the breakthrough technology to smooth the way for methanol in a decade. Who knows, we might end up becoming the holding company for this thrust instead, for the DMFC for portable electronics is already close at hand to replace batteries.

This is heresy, but my interest is not in necessarily making a pile of money. The more important objective is to initiate the development for what many of us believe to be the most sensible biofuel: methanol. I look forward to your comments.

Aloha.

Pat

Comments (2): Not only was there a dearth of comments, the letter itself attracted no new interest. Well, I tried.

My next article published on 14June08 made fun of superstitions and large numbers.

Piffle Squared[22]

Piffle squared? That's what we think of the national renewable energy program.

What is the relationship between Friday the 13th and Father's Day, both which occurred this weekend? Nothing much, although this coincidence provides me yet another opportunity to point out the wisdom of priorities we have on how we spend our money.

Some of you suffer from paraskevidekatriaphobia, a fear of Friday the 13th. Buildings sometimes skip that floor and people tend to be a bit more careful that day. Almost a billion dollars is said to be lost to business on this day and the British Medical Journal reported that there is a significant increase in traffic accidents on that day.

You can probably blame the Bible for most of this nonsense because Eve offered that fateful apple to Adam, Christ was crucified and Noah's Great Flood began, all on Friday the 13th. Judas Iscariot was the 13th guest at the Last Supper. This will be the only Friday the 13th in 2008, but three are coming in 2009.

As Christianity is not popular in the Middle East or Asia, they don't bother worrying over this day over there. However, the number 4 is considered to be bad luck in Japan, so don't give any presents of four pieces. Why? Because the number 4 is pronounced shi, which is also the word for death. Superstitions are generally based on such things.

We also have Father's Day in the U.S., which is celebrated world wide throughout the calendar year. Our first Father's Day occurred almost exactly a century ago, on July 5 in 1908 in Fairmont, West Virginia to commemorate a coal mining tragedy.

President Woodrow Wilson was instrumental in making this day popular, but it was President Lyndon Johnson who made this a Sunday "holiday," officially recognized by President Richard Nixon in 1972 on the third Sunday in June.

With Mother's Day, these remembrances help keep business in business. Sons and daughters spend $30 billion each year to honor their parents. This is more than thirty times what the U.S. Department of Energy (our tax money) annually sets aside for renewable energy research. Hmm, maybe one year we should apply all the money spent on these two celebrations towards saving Mother Earth and prolonging the life of Father Time, perhaps today also known as Peak Oil.

---

22 http://www.huffingtonpost.com/patrick-takahashi/piffle-squared_b_107137.html

Oh, yes, we treat Peak Oil / Global warming about the same as Friday the 13th: we're somewhat careful, but hope that nothing bad will happen. As a Nation where 90% of us believe that there will be some kind of afterlife, it is no wonder that we can blithely go on hoping for the best, as underscored in *SIMPLE SOLUTIONS for Humanity*.

Yes, $30 billion is a piffle compared to the $45 trillion reported by the International Agency as needed to just deal with mitigating climate change. But, again, $30 billion is more than thirty times what we currently spend on renewable energy research. Thus, Piffle Squared!

Comments (0): No comments. Hmm, I guess humor is taken silently.

Over time I noticed that Republicans supported their funding sources (big business) and Democrats took care of the people. You would think just this difference would insure for one party rule, for there are so few rich people. Ah, but politics are not that simple.

Why Do Republicans Like Fossil Fuels and Not Care That Much for the Environment?[23]

Well, a lot more oil/coal money is given to Republican candidates, who therefore display their loyalty when elected by voting for this greenhouse gas energy source and against the environment. While that might be too simplistic an explanation, a good example of such action is Republican Congressman and Whip Roy Blunt (R-Missouri), who, on June 5, 2008, displayed on his web page the following:

	REPUBLICANS	DEMOCRATS
ANWR* Exploration	91% supported	86% opposed
Coal to liquid projects	97% supported	78% opposed
Oil shale exploration	90% supported	86% opposed
Offshore oil exploration	81% supported	83% opposed

*Arctic National Wildlife Refuge

He was merely pointing out that Republicans have been trying to develop homegrown energy reserves while Democrats were the ones responsible for our current energy predicament. Of course, it is no surprise that the League of Conservation Voters, a nonpartisan environmental group, gave him the lowest possible score, zero, seven of the past eight congressional sessions.

Last week in the U.S. Senate, Democrats could not muster the votes to overcome a Republican filibuster on a plan for a windfall profits tax on the oil industry. The week before that Republican Senators killed the climate change mitigation bill. I loved Senator Bernard Sanders' quoted response: "The American people must be wondering what in God's name is going on in their nation's capital." Sanders is an Independent from Vermont.

When I worked in the U.S. Senate, President Jimmy Carter had a progressive solar energy program. But that was mostly because we were in the depths of the second energy crisis and there were such things as gasoline lines. I was still there when Ronald Reagan became president and decimated the national solar program. I went on to become director of the Hawaii Natural Energy Institute in the mid-80's when there were very little Federal funds for renewable energy research. Hub Hubbard, director of the then Solar Energy Research

---

23 http://www.huffingtonpost.com/patrick-takahashi/why-do-republicans-like-f_b_108190.html

19

Institute (now National Renewable Energy Laboratory), and I had an inside joke of not getting much funding, but were nevertheless increasing our market share to keep surviving. ***SIMPLE SOLUTIONS for Planet Earth*** goes into the politics of renewable energy.

Mind you, the only Democratic president since then, Bill Clinton, did not do very much for sustainable resources. In fact, somewhat influenced by the aura of global warming, his administration gave their full blessing for increasing natural gas use over coal for electricity production. So what happened? The U.S., in seven years, added more gas-fired electric generation capability than the entire capacity of Europe. So what happened? The price of natural gas tripled, and it is said that not only did American consumers end up paying a lot more for energy, but we also exacerbated our carbon footprint. We have not been very smart in our energy planning, which can be expected, for we still don't have any national energy policy (second essay).

The ***Congressional Insiders Poll*** reported on 7June2008 showed that 95% of Democrats and only 26% of Republicans agreed with the following statement: "Do you think it's been proven beyond a reasonable doubt that the Earth is warming because of man-made pollution?"[24]

John McCain on 18June2008, called for the construction of 45 new nuclear power plants by 2030, drilling for oil in the coastal zone and, again, a summer waiver of the Federal gasoline tax. Barack Obama disagrees on all the above and has continued to emphasize conservation, mass transit and wind, solar and green energy. McCain co-sponsored the climate change mitigation legislation in the Senate, so Obama will not be able to make this major Republican-Democrat difference an issue. But the evidence is overwhelming that Republicans love those fossil fuels.

How are these inclinations affecting their electability? In a ***CNN/Opinion Research Corporation Poll*** of 4-5June 2008, voters nationwide were asked, "Which of the following issues will be MOST important to you when you decide how to vote for the president? The results were: the economy (42%), war in Iraq (24%), health care (12%), terrorism (11%), illegal immigration (8%) and other (1%). Peak Oil and Global Warming, thus, combined, rate less than 1%, and did not even make the list.

A **Los Angeles Times** / ***Bloomberg*** Poll of May 1-8, 2008, essentially asked this same question and the environment got 4%. No mention of gasoline or energy. Thus, are Republican/Democratic platforms on energy and the environment irrelevant?

Actually, no, as the ***Pew Research Center*** survey of May 21-25, 2008 showed that registered voters thought energy (77%) and the environment (62%) were important issues. Why this remarkable discrepancy among polls? Mainly, the other surveys asked what was the one MOST important issue.

Gasoline sells for more than $10/gallon in Europe (in the $11/gallon range in Germany), so even $5/gallon in the U.S. should be a blessing. Well, truckers will begin to find a way to rebel,

---

24 http://nationaljournal.com

lifestyles will be compromised and the matter of energy as a problem will grow through the summer, especially if Morgan Stanley is correct that oil will hit $150/barrel by Independence Day.

So what is the conclusion? A wild card in this upcoming election is that McCain and Obama are at polar opposites on the matter of energy. The issue of fossil / nuclear energy versus green energy could well be the uniquely different determining factor in November. We of course need both to minimize the trauma of Peak Oil and Global Warming, but the voting public will actually be able to make a difference in the selection of the 44th President of the United States, and the margin of victory could well be green.

Comments (40): Well, the title of this posting struck a nerve. Some were insulted and many were feisty. Good! I enjoyed the exchanges.

On 24June08 came simple solutions to Peak Oil and climate change. We are so flummoxed by politics and conflicts that the result is nothing. So, let's start with something simple, and work at it.

Simple Solutions to Energy and Global Warming[25]

The June 19th issue of *The Economist* very well captured the state of energy and global warming today, with vague hints on future directions. Some consensus, though, appears to now be forming about the current situation and potential pathways:

1. The global economy is unsustainable and possibly headed towards collapse.

2. Peak Oil plus surging demand will continue to increase the price of petroleum.

3. Global Warming will result in a Carbon Tax.

4. It will take improvements in energy efficiency and a range of sustainable energy resources to minimize the coming pain, and nuclear power will no doubt be included.

The problem is that society suffers from a fatal flaw: we can't seem to make smart decisions until it is too late. Thus, we will agonize, then suffer, but, somehow, again recover. Is there a better way?

We should have done something after the first energy crisis in 1973, but did not even do anything much after the second energy crisis in 1979. Part of the problem is that Ronald Reagan decimated the American solar program when he became president in 1982, and the world followed suit. Understandably, too, there was no need for panic, as gasoline (in 2005 dollars) cost about $2/gallon in 1950, jumped to $3/gallon in 1980, but dropped to $1.30 in 1998. (Go to *SIMPLE SOLUTIONS for Planet Earth* for details on this paragraph and much of the following.)

Thus, the 1997 Kyoto Protocol on climate change came at the worst time possible when oil prices were nearing an all-time LOW. Again, in 2005 dollars, oil cost $18.76/barrel in 1972, $89.48/bbl in 1980, $14.38/bbl in 1998... and is now somewhere over the rainbow in the range of $135/bbl, in 2008 dollars. The fact that China and India were excused made it easy for the incoming George W. Bush administration to also opt out in 2000. No matter how conscientious you want to be, reducing carbon dioxide became a joke when those countries conforming to those environmental rules began to lose investments and jobs, made all the more hilarious when just the additional coal power plants of China, the U.S. and India dwarfed all the carbon dioxide savings accrued by the 182 parties who signed and took steps. By the way, India and China did ratify the protocol. It's just that they don't have to follow the strictures

---

25 http://www.huffingtonpost.com/patrick-takahashi/simple-solutions-to-energ_b_108783.html

In 2006 came the 600-page Stern report[26] commissioned by British Chancellor Gordon Brown. The solution was simple: annually invest 1% of the World Gross Domestic Product to stabilize greenhouse gas concentration over the next 50 years to prevent a global recession. In 2007 the Intergovernmental Panel on Climate Change said we might need to invest as much as 3% of the GDP by 2030 to limit long-term global warming. This is serious money, as a household in the UK with a weekly income of $700/month would be taxed an additional $91/month (or $3 per day) for this program.

Then on June 6 of this year, the International Energy Agency indicated that the world would need to spend $45 trillion[27] by 2050 just to meet the target of a 50% cut in emissions suggested by the G8 nations, where the atmospheric temperature would still probably increase by a little more than 4 degrees Fahrenheit. The 2008 U.S. Federal budget is $2.66 trillion, so we are talking really big money, and not even getting close to solving the problem.

International study groups can suggest all they want, but someone has to make a decision. The G8 nations next meet from July 7-9 in Japan, where the main theme is the environment. Will there finally be a firm resolution to Peak Oil and Global Warming? No! No! No! No! No!

Our leaders will dialogue and express sincere concerns, but we might need to wait until 2012 when those heads of countries meet in the United States, when someone like, perhaps, Barack Obama, will be in his 4th year of presidency. It will take all that time, anyway, to gain some consensus and arrive at a workable plan. But actual decisions will be made only if the world is by then in the midst of depression, where, maybe millions have succumbed to a particularly hot summer, and oil rests at $250/barrel. But, then, isn't that a bit too late?

So what is the SIMPLE SOLUTION? Read my ***SIMPLE SOLUTIONS*** books. The first action step I took was reported in my first HuffPost of 29May2008, entitled, "Well, Barack, We Have a Problem."[28] But what was I thinking when I only suggested only a trillion dollar solution.

Comments (8): Some agreed and some didn't on whether solutions can be simple. I'll say it again, the solutions are simple, but the process to get there could be impossible.

---

26   http://www.abc.net.au/worldtoday/content/2006/s1776868.htm

27   http://www.reuters.com/article/idUSL0348323120080603

28   http://www.huffingtonpost.com/patrick-takahashi/well-barack-we-have-a-pro_b_104201.html

The notion of geoengineering at a global scale scares most. Their argument is, we have so screwed up our planet that to entrust any group to solve these problems at mega-scale is insane. They have a point, but the concept merely suggests we talk about these super solutions. My 1July08 posting reviewed these early thinking steps.

Geoengineering of Climate Change[29]

There are those who feel that there is no cause for concern at this time about the increasing amount of carbon dioxide in our atmosphere, and if you trace who they are -- as for example by perusing through the comments of postings like this -- organizations like the Advancement of Sound Science Center or the Heartland Institute seem to regularly pop up. Searching further, you see that companies like Exxon Mobil provide supporting funds. Our White House provides encouragement and Republicans more than Democrats side with these detractors.

For all I know, they might actually be right. However, let's, for the sake of discussion, say that global heating is real and our world leaders are unable to agree on a workable solution in time. What if the situation gets so bad that virtually instant solutions will be required to save our civilization? I provide a wide variety of answers in **SIMPLE SOLUTIONS *for Planet Earth***, but for the purpose of this article, let us look at something called global geoengineering.

Various international conferences on this subject have been held over the past few decades, but, in general, proponents have generally been relegated to the lowest level of respectability by academics and funding agencies. Until, maybe now.

The concept is not new. The industrial revolution, farms, cities, transport systems and remedying the ozone hole can be considered to be forms of geoengineering. The Montreal Protocol actually seems to be working for the latter, but the Kyoto Protocol has been less than successful.

How can you quickly reverse global warming? It has been hypothesized that reducing sunlight by only 1% should eliminate this problem. Various ideas have been floated, from placing reflective sheets on the ocean or in space to exploding a controlled series of hydrogen bombs to stimulate a nuclear winter. Yes, some of the propositions have been certifiably insane.

One I favor (see the chapter on the Blue Revolution in the book mentioned above) has to do with an Apollo Project equivalent of building an armada of open ocean grazing platforms powered by ocean thermal energy conversion to suck up carbon dioxide from the atmosphere while providing new habitats, green materials, next generation fisheries and sustainable fuels. Alas, such an effort will take decades and, horrors, maybe result in a United Nations of a thousand members.

---

29 http://www.huffingtonpost.com/patrick-takahashi/geoengineering-of-climate_b_110122.html

The concept that has gained the greatest traction is the stratospheric sulfate solution (S-cubed), where large amounts of sulfur dioxide are, through various mechanisms, placed at altitude. This gas would form droplets of sulfuric acid in stratocumulus clouds to reflect back sunlight into space. Names like Freeman Dyson, Paul Crutzen and Edward Teller appear as advocates. This cure might cost $100 billion/year, for the effect wears out after a year, but that is a piffle in comparison with the $45 trillion exclaimed by the International Energy Agency as necessary to insure that our surface temperature only increases by 4 degrees Fahrenheit by the turn of the century.

Surely enough, Mount Pinatubo in 1991 blew its top and threw 20 million tons of sulfur dioxide into the stratosphere and the globe cooled about a degree Fahrenheit that year. So the basic S-cubed concept has been largely verified by nature.

Before anyone gets too irrational, let me underscore that no one, not even the most extreme supporter, is even suggesting that anything of any magnitude be initiated today. It wouldn't hurt, though, to set aside a small amount, perhaps 1% of the global change budget, to comprehensively study the more reasonable suggestions, especially reviewing the environmental implications, so that if that one in a hundred chance that a perfect global heating storm (as, perchance, depicted in The Venus Syndrome chapter of **SIMPLE SOLUTIONS for Planet Earth**) actually happens, we will have a few rational emergency options worthy of consideration.

Comments (4): I thought this issue would have drawn a lot more irate responses, but those that did provide input either missed the point or supported my "insurance" strategy.

Every so often, I'm compelled to reach outside the box for a solution. On 2July08 I extemporaneously blurbed at an awards ceremony to make hydrogen free. Having said so, I now felt responsible to follow up. In four parts I morphed through various stages of belief and disbelief about this notion.

What About Free Hydrogen? (Part 1)[30]

Over the next couple of Green posts I will discuss the matter of free hydrogen. Yes, just make hydrogen free by, say, 2020, and let industry, with government assistance, develop the infrastructure and systems to handle the Age of Free Hydrogen. At first glance, the concept appears insane. For one, many (mainly government officials, actually) of the responses I received to this suggestion expressed concern that energy use would get out of hand, for then no one would conserve. Well, maybe that might actually be okay, but, clearly, the matter is complex. The details will need to be well thought out. For example, the hydrogen, in this context, must, of course, come from renewable energy. Then, will it be possible to differentiate between cheaper dirty hydrogen and the more expensive clean hydrogen? Also, who will be providing this free energy? As to be discussed, you will, the taxpayer. However, this simple solution should ultimately be able to eliminate the negative repurcussions of Peak Oil and onset of Global Warming. Interestingly enough, over the past year since the book was published, I began to appreciate the value of, possibly, a more logical sustainable pathway, to be revealed in the final post of this series. Anyway, the following is from Chapter 3 **of SIMPLE SOLUTIONS for Planet Earth.**

On March 21, 2006, at the annual luncheon of the National Hydrogen Association (NHA) Conference in Long Beach, California, I received the Spark Matsunaga Memorial Hydrogen Award, usually given to an elected official. However, as I was the individual who U.S. Senator Spark Matsunaga assigned in 1980 to write the first draft of his hydrogen bill, I guess I was considered to be close enough to qualify. The second recipient, in 1992, was U.S. Senator Daniel Akaka, whose letter of congratulations was read by Jeff Serfass of NHA. Other awardees have included Congressmen and Senators, although Governor Arnold Schwarzenegger received this honor in 2004. Walking up to the podium, aside from the assorted obligatory thank you, I wondered what I was going to say. (Pardon me for mentioning all this, but a degree of credibility helps when one leaps beyond the edge of the envelope.) It then came to me in a moment of splendid inspiration, bursting forth from a third of a century of deliberation-- MAKE HYDROGEN FREE. Deep in my memory might well have been a statement by Jeremy Rifkin in his book on **The Hydrogen Economy**, where he imagines a future a century away where the cost of producing unlimited amounts of hydrogen should virtually be zero. This sounds too much like atomic power being too cheap to monitor, but let me proceed.

---

30 http://www.huffingtonpost.com/patrick-takahashi/what-about-free-hydrogen_b_110532.html

Some say that hydrogen will always be a bit, if not a lot, too expensive. Then, too, this is a chicken or egg problem. The dilemma is in the infrastructure and free market system for what is an artificial commodity. Where do you start?

As great as, say, clean hydrogen sounds to some romantics, you can't force this gas on society just because it seems to them so logically sensible as the universal fuel of choice. If mankind is, indeed, at a decisive juncture, a means must be found to more effectively induce the world to quickly transition from a fossil economy to something better.

We'll come to the how later, but an ideal alternative worthy of discussion would be one powered by clean and sustainable hydrogen. The fuel would be produced everywhere. There would be no OPEC, no nuclear terrorism, and only a vibrant and healthy Planet Earth. World Wars would be minimized because most of the big ones, including the current action in Iraq, were fought over limited resources.

WHY NOT CONTROL THE ISSUE BY MAKING HYDROGEN FREE? What a heck of a simple solution for energy and our environment! If the perfect vision of 2020 is not possible, push the operational date back a bit. If the crisis is upon us, do it now. As opposed to waiting for economies of scale reducing the price to a competitive level in a century, start with zero and keep it there. There will be a transition period for the hardware to become available, but there are ways to administrate this process, and wouldn't it be wonderful if it works? The Free and Clean Hydrogen Age would eliminate our growing climate warming problem, while going a long ways towards preventing world wars forever and enable our civilization to hurdle over the Peak Oil problem.

(To be continued.)

Comments (32): This article also drew respondents out of the woodwork. I thought input would be provided after all four parts were posted, but, no, many wanted to comment now. As a result, some of the feedback was premature or irrational. A few got the idea right, but let's see how the series evolves.

*On 3July08 I described the follow-up to my Hydrogen Man of the Year statement. To make a long story short, no one mentioned bothered to even respond to my entreaties. I exchanged various e-mails with the Board of the **Journal of Hydrogen Energy** (all close friends) and the most supportive was from the editor who indicated that I should first do the research before submitting a paper for publication. I told them I did the research, and the facts reinforce my bluster.*

What About Free Hydrogen? (Part 2)

*The following continues the discussion on a free and clean hydrogen age, as largely excerpted from Chapter 3 of **SIMPLE SOUTIONS for Planet Earth**.*

Referring back to that National Hydrogen Association awards luncheon, the fact that I had the courage or idiocy to introduce this concept before giving any thought to it was, on hindsight, foolhardy. But, if I had cogitated over it, nothing would have happened. There is something unpredictable and serendipitous about such acts.

This extemporaneous message was suggested to the thousand or so in the audience. At lunch I sat next to Phillip Baxley of the Shell Hydrogen group, who was the incoming chairman of the board of NHA. The fact that I was one of the original organizers of this association and a past board member, I thought, would carry some credibility. I was particularly interested in how he would react because, on the surface, oil companies would seem to have the most to lose. In reality, Peak Oil is changing the topography of energy economics, and petroleum suppliers must start today to make monumental corporate decisions on what they will become tomorrow. A Free Hydrogen Age actually provides them a SIMPLE SOLUTION. I asked Phil to think about this for a year and later share with me his reasoned reaction. A year henceforth, still no feedback. (*Still no response today, more than two years later.*)

However, the luncheon setting was not conducive to questions and answers, nor comments. So, two days later, at the breakfast gathering of a fairly recently formed International Partnership for Hydrogen Energy (IPHE), I waited till the end of the briefing and discussion to again toss out the concept of what I then termed the Free Renewable Hydrogen Age. In a nutshell, I asked the stakeholders of IPHE to discuss this concept and, if they determined that the idea was patently ridiculous, please assign someone to e-mail me why. I received wide applause from those in attendance and the meeting ended. This reaction emboldened me to press forth. So I quickly drafted a Free Hydrogen paper and sent it to Graham Pugh of the U.S. Department of Energy and Thorsteinn Sigfusson from Iceland, who were co-chairs for the discussion, and asked for their impressions. More than a year later, no response. (*Still none in mid 2008…and none in 2010.*)

The IPHE is ideal for this purpose because it is relatively new, has representatives from the U.S., European Community, Russia, China, India, Japan and other countries -- representing 85%

of the world's Gross Domestic Product -- and meets several times each year. I noted, though, that their objectives and goals looked similar to sundry other international agreements of the United Nations and International Energy Agency, and they were already at risk of being bogged down by politics, protocol and assorted conflicts. I have personally been involved with a whole range of these types of activities, and they all just sort of turned into mush. Great for travel, mind you, but essentially ineffective and a huge waste of time. What they needed to do, I suggested, was forget about evolutionary developments and trying to appease everyone, but, instead, take a revolutionary approach to make a real difference. Start with the possibility of free hydrogen, or something grandly similar, and plot out an optimal strategy to get there.

Such total disregard from experts in the field I guess must mean that they don't think much of the idea, and, worse, might signify that it is so beyond the pale that any reply is unworthy of their status. On the other hand, perhaps, they could not find fault and are paralyzed in their process of analysis. More probably, they are all too busy doing more important things. My sense is that the Free and Clean Hydrogen Age is but one of a million solutions with potential, and I remain in reasonable standing among my peers, for the ***International Journal of Hydrogen Energy*** has not yet kicked me off their editorial board.

*(If you want the scientific validity of the Hydrogen Economy, go to **SIMPLE SOLUTIONS for Planet Earth**, Chapter 3. Parts 3 and 4 of this continuing series will provide some economic realities, and, perhaps, a better solution.)*

Comments (0): Well, to my surprise, again, there was no response. So let's go on to Part 3.

Well, came 4July08 and I felt like using this day for energy, so delayed my Part 3 of Free Hydrogen. Crude oil had reached $145/barrel, and I thought the timing was ideal to catalyze a crusade for energy independence. Let me take advantage of the fact that this posting was being read by thousands and the world-wide web was an ideal medium to foment a rebellion.

We Need a Declaration of Energy Independence[31]

On this 4th day of July, 2008, when oil rests at its highest price on record (just above $145/barrel), the time has come for a Declaration of Energy Independence. Government seems unable to take any meaningful action, so, maybe the world wide web can be the action ground. Let us use something like a reverse Ponzi scheme, where each person reading this post sends it out to ten friends, who in turn... No, nothing terrible will happen to you if you don't, unless you count the double hammer of Peak Oil and Global Warming as the implied enforcer.

According to http://www.CleanEnergyAction.net, Americans believe thusly:

- 90% feel that our federal government is not doing enough about energy,
- 82% say that the U.S. should be a leader, not a follower, on global climate mitigation, and
- 75% believe that there should be a 5-year moratorium on new coal-fired power plants.

Mind you, the world has for the longest time now been concerned about energy, and limited resources spurred both world wars and our current presence in the Middle East. Some give credit to Richard Nixon for first mentioning the concept, when, following the First Energy Crisis in the Fall of 1973, he announced Project (Energy) Independence in his 1974 State of the Union address. Jimmy Carter, in 1977, promulgated his unfortunate Moral Equivalent of War (MEOW) plan to attain energy self-sufficiency by 1990. Then came the Second Energy Crisis in 1979 and renewable energy funding jumped to more than $2 billion (in 2005 dollars) by 1982. That was when the world should have initiated the so called Manhattan project for sustainable resources.

Unfortunately, Ronald Reagan was inaugurated in 1982 and decimated the solar program. Amazingly enough, the Department of Energy alternative energy budget since then never reached even $1 billion/YEAR again. Considering that Americans spend about a billion dollars each DAY on gasoline, a Nobel Laureate economist reports that we expend a billion dollars each DAY for our war in the Middle East and each space shuttle shot costs about a billion dollars, one can only be embarrassed by our chosen priorities.

The primary reason, of course, was that, in 2005 dollars, oil dropped to $14.58/barrel, when it was as high as $89.48 in 1980. In 1972, before that initial energy crisis, oil cost $18.76 (yes, around $3 those days, but we are talking 2005 dollars).

---

31   http://www.huffingtonpost.com/patrick-takahashi/we-need-a-declaration-of_b_110948.html

Mind you, virtually every pundit since Watergate has continued to pound on the myth of energy independence. Ah, but our U.S. Congress last year did actually pass the Energy Independence and Security Act, which did not do much of anything. Perhaps in response, Robert Bryce earlier this year published Gusher of Lies: The Dangerous Delusion of Energy Independence. So, be careful, for anyone supporting anything resembling energy independence could well be forever discredited.

So, sure, vote for a president to, for the first time, establish a national energy policy. Yes, make sure our elected representatives more seriously consider something like a carbon tax to reduce global warming. Are you dreaming? If nothing progressive has happened for all these past few decades, what gives you any hope for a sudden attitudinal change? Unless, that is, you do something about it.

I have kept saying that there is something about these blogs, featuring instant feedback, that can make that crucial difference. Let us pronounce, today, our Declaration of Energy Independence, and send this proclamation to ten of our closest colleagues, asking them to do the same, or, at the least, link this post with another blog site. Will this work? Nothing else has.

Comments (8): Alas, there was no virtual revolution. The response was generally supportive, but a couple asked what happened to Part 3. A real disappointment. I think I need to link directly with Twitter, Facebook and all those other portals.

After my public challenge about Free Hydrogen, I did some research and made some calculations, as shown below in my 8July08 posting.

The Free and Clean Hydrogen Age (Part 3)[32]

This discussion continues the concept of Free Hydrogen

Many of us demand free education through high school, good roads, a functioning military, public safety security, etc. Michael Moore's **Sicko** reveals that even medical care is largely free in Europe and rest of the world. We pay for all of this, of course, through taxes. Is there something universal about a similar energy entitlement? We get very close with electricity, anyway.

SIMPLE SOLUTIONS for Planet Earth provides most of the details in Chapter 3, but, to get to the current bottom line, the average family in the coming years will annually spend about $5000 on gasoline and pay $25,000 in taxes. It is possible that free hydrogen can be supplied if this household provides this same total, that is, $30,000 in taxes each year. The transition will be messy, and the devil will no doubt be in the details, but free transport fuel with the added benefit of a cleaner environment and more secure world would be worth it, even if a few extra dollars (in the form of a carbon tax) might initially be required. Ignore the semantics, call this an investment for your future and that of Planet Earth.

The total analysis, would, of course, need to include all energy factors, so let us look at the entire picture. Over the coming years, there will be more than 120 million households on average spending $8000 for powering ground transport, heating and electricity, or nearly a trillion dollars each year for energy. A 10 cent/pound carbon dioxide tax would increase gasoline prices by $2/gallon (even with this increase, the domestic cost at the pump will still be only HALF what is paid in Europe today) and coal-fired electricity by 8 cents/kWh. Note that under those conditions, wind power, solar thermal and future paint on solar photovoltaics can compete against coal-fired electricity. There is that added matter of nuclear power that needs to be considered in this mix. This extra carbon investment revenue to be generated could, thus, in addition, be more than a trillion dollars/year. This grand total of $2 trillion/year can thus be applied towards developing a free hydrogen economy, as outlined in **SIMPLE SOLUTIONS for Planet Earth**. In time, the hydrogen jetliner, too, can be accommodated.

This is just a gross calculation, but the point of all these numbers is that government won't need to suffocate the economy to produce sufficient revenues to support a free and clean hydrogen economy. Remember, according to Nobel Laureate Joseph Stiglitz, just the Middle East war will cost you, the taxpayer, at least $3 trillion.

---

32   http://www.huffingtonpost.com/patrick-takahashi/the-free-and-clean-hydrog_b_111323.html

The role of industry will be as an equal associate, possibly in government-company partnerships. Ultimately, by 2020, or certainly by 2050, when the technologies/infrastructure becomes available for hydrogen, either government will merely tax you an amount that you would otherwise be paying for energy anyway, partially subsidized by the savings to accrue from a lower defense budget (see HuffPo on "Well, Barack, We have a Problem..."), or a means will be found to continue the one-to-one relationship between the consumer and public utilities/energy sector. We thus can hopefully sidestep Peak Oil, remediate global warming, circumvent a global economic depression and minimize future world wars.

There is, of course, one huge problem with this analysis. We could be faced with these environmental and economic cataclysms today, or very soon, while it will take decades for a hydrogen infrastructure to develop, even under emergency conditions. Thus, while we must act now (and we really should have in 1980 during the aftermath of the second energy crisis), there, unfortunately, is a Catch 22 dilemma -- a free hydrogen solution, even if deemed optimal, can only be invoked AFTER any mega crisis. That's the reality of our democratic free-enterprise society. How, then, faced with this fatal flaw in our civilizational make-up, do we implement this plan? No, a benevolent dictatorship is unlikely. Maybe the answer is not hydrogen, or perhaps we are doomed, but Part 4 is yet to come.

(This is #3 of a 4-part series.)

Comments (2): There were only two comments, but Part 4 is yet to come.

Part 4 somehow got lost in the Huffington Post system. The following came from my daily blog of 17July08.

The Free Hydrogen and Electricity Age (Part 4)[33]

To recap parts 1 to 3, the FREE Hydrogen Age was advocated as a solution to be considered if the world got suddenly clobbered by the combined hammer of Peak Oil and Global Warming, triggering a global depression. Certainly, under business as usual conditions, the notion of FREE hydrogen, or FREE anything, would only draw ridicule. These posts were also carried in The Huffington Post (http://HuffingtonPost.com).
-

Last week, at the G8 meeting in Japan, our global leaders made what to some might seem like a promising declaration: cut carbon dioxide by half by 2050. The problem is that this is kind of what they also said last year in Germany, except, this time with weakened language. Is this progress? In a sense, one shoe (the environmental one) has, thus, fallen.

Regarding the other, Peak Oil and prices, former Shell chairman, Lord Ron Oxburgh, warned in September of 2007, that oil could reach $150/barrel, and in November, Usameh Jamali of OPEC said the same. Morgan Stanley was more specific, and earlier this spring predicted $150/barrel oil by July 4. Well, that did not happen, but, at $146/barrel, got awfully close the day prior. In May, Goldman Sachs forecast $200/barrel oil by the end of the year. A really hot summer, where millions succumb, plus oil at this lofty range, could well trigger a devastating economic plunge to set the stage for that FREE Energy Age.

I recall an *American Scientist* cartoon[34] way back when where a professor at a blackboard solved a difficult problem by inserting "a miracle happens." Avoiding the hard question of who will actually make the command decision (I think the G8 group should be it, but that is another posting) and how, let's say a miracle, in fact, occurs. A legal proclamation is made to make hydrogen free by January 1, 2020. How might the transition look?

First, it will be extremely difficult to provide unlimited free hydrogen by that date, but not impossible. The infrastructure is currently lacking. You can't instantly convert the ground and air transport system to use hydrogen. But that's not the point, for industry will do all it can and begin maximizing the availability of anything that uses hydrogen. With wind power and all the other solar options, made competitive by a severe carbon tax, facilities can be mass-produced to make free hydrogen. The supply should at least match the means to utilize it. Would electricity, too, be made free? Something to consider if generated from a renewable source. So, maybe we should be saying, FREE Renewable Hydrogen and Electricity Age. In any case, if everything works to perfection, only a relatively small fraction of actual

---

33 http://planetearthandhumanity.blogspot.com/2008/07/to-recap-parts-1-to-3-free-hydrogen-age.html

34 http://www.amazon.com/Whats-Funny-About-Science-Scientist/dp/0913232394

energy utilized in 2020 will in fact be FREE hydrogen, or electricity from anything related to hydrogen, but this is of secondary importance, for an exponential trend will have been initiated.

The Free Hydrogen Age will need a bridging renewable liquid fuel, and the world, as I intimated in an earlier post ("What is the Best Biofuel?"), seems headed down a dead end bioethanol / biodiesel pathway. Either go to Part 2 of that article ("Ethanol versus Methanol") or Chapter 2 of my book on SIMPLE SOLUTIONS for Planet Earth and learn why biomethanol is the wise choice.

-

Part 4 is now getting too long, so let me stop here. For Part 5, you can today go to http:// huffingtonpost.com/, type "Patrick Takahashi" into the upper right box, click on SEARCH, than click on Part 5, which will also be covered in this blog tomorrow.

Comments (0): No one comments on my daily blog.

Well, on 17July08 I published Part 5 of the series, and the final version is entitled not Free Hydrogen, but Free Green Energy. I was influenced by many that hydrogen, per se, might not be the ideal fuel, but an assortment of clean options could well be. I agreed.

The Free Green Energy Age (Part 5)[35]

This is the fifth and final article of my HuffPo series on finding an optimal solution for Peak Oil and Global Warming. There is a clear transition of thought from Parts 1 to 5. These points are covered in **SIMPLE SOLUTIONS** *for Planet Earth*.

Did you know that we will, over the next year, send to foreign oil producers around $700 billion to pay our annual oil bill, while we invest less than $1 billion on renewable energy research? What are our personal priorities? Americans annually spend $25 billion on video games, $80 billion on cigarettes and $100 billion on alcohol, with a huge subsequent downside on time lost, health and relationships. Problem #1: the lack of public will regarding critical national priorities.

Further, the U.S. and Venezuela reached Peak Oil in 1970, Iran 1974, Indonesia 1991, Norway 2000, Mexico 2003, and Russia 2007. Kuwait, Iraq and Saudi Arabia will all reach this pinnacle around 2015. British Petroleum reported in 2000 that the world oil production per capita crested in 1979.

It would thus not particularly surprise me if we later learn that Peak Oil occurred in July of 2008, or earlier. A telling sign is that the price of petroleum last week exceeded $147/barrel, the highest ever, then crashed $10 on Tuesday, showing a dangerous metastable predilection. The Dow Jones Industrials reacted by dipping below 11,000, now officially a bear market, dropping from the 14,165 in October. What will happen when oil reaches $200/barrel?

The International Energy Agency reported on a $45 trillion bill just to maintain our current level of greenhouse gases into 2050. The G8 leaders last week in Japan ho hummed by again passing the buck.

Part of the problem with global warming is that people don't get too excited about a tenth of a degree temperature increase each year or a small fraction of an inch sea level rise, unless you happen to live on an atoll. In my two posts on The Venus Syndrome, though, I discussed the prospects of a terminal cascade effect leading to 900 degrees Fahrenheit. Also too, to feed your future nightmares, if all the ice melts, the ocean will elevate more than 250 feet.

While these worst case scenarios are sufficiently distant in the future to worry us, a few enterprising spirits are beginning to get concerned. T. Boone Pickens, a Republican and oil

---

35   http://www.huffingtonpost.com/patrick-takahashi/the-free-green-energy-age_b_112967.html

man, proposed a trillion dollar wind farm effort. Politicians from various quarters are now suggesting a range of next generation Manhattan Projects. Problem #2, yet, is politics, itself. Another T., Booker Washington, once said "there are two ways of exerting one's strength: one is pushing down, and the other is pulling up." Today, our elected officials mainly concentrate on the former. Let's stop beating around the, ahem, Bush. Instead, focus on the latter, and, while we're at it, do something monumental, like making *green energy free* some time in the future, say 2020. James Hanson of NASA leaned in this direction when he advocated a severe carbon tax, but wanted the revenues directed back to the taxpayers, not government.

So, after four parts, the punch line is the *simple plan: eliminate all energy incentives, even for renewables...forget about the carbon tax...just make sustainable and clean energy in 2020.* A dozen years is close enough to be meaningful, plus 2020 is symbolically representative of good vision. We really should have initiated that energy Apollo Project after the second crisis in 1979, for now, the doom projected by some, is worrisomely close at hand.

Some might exclaim, what about those poor coal miners in West Virginia, or nuclear plant operators or economies of the Middle East oil producers? Not to worry, for wind, solar and bio represent on the order of 1% of our supply today, and traditional energy forms will be required for decades to come. What will, for example, petroleum cost? Don't know, but the market will determine prices into and beyond 2020. It's just that Green Energy will be supplied for free from that year. Who will provide this energy? Not sure, probably some government-company partnership. After all, today, electric utilities are already closely controlled, so this is not new. The primary benefit of this epic program will be to bring safe and home-produced energy to the consumer as soon as possible to minimize the economic and lifestyle trauma surely to come anyway. You see, we missed the boat in 1979 and will thus suffer some consequences.

A universal *free* green energy declaration will provide incredible opportunities to creative people and our free enterprise system. If you have been reading my various **HuffPos**, I have said that ethanol from fermentation will probably be replaced by methanol from gasification/catalysis; the direct methanol fuel cell will soon be readied to supplant internal combustion engines and batteries; wind power will truly surge; and hydrogen jetliners will be flying and fusion power will now be commercialized long before 2100.

But traditionalists will state, this is impossible...in fact, crazy. So be it. I provide a hint in my first **HuffPo** of May 29 entitled, "Well, Barack, We have a Problem..." Then through all my other posts, leading to this series, details about purposefully controversial alternatives are provided. I do suggest that the G8 Nations and United Nations take the lead, but, on afterthought, the USA made an early unilateral decision to legislate for clean air and water, and the world followed. Our next president, Congress and the private sector must set aside their differences, ala Booker T, and take just one magnificent step: make Green Energy *free* in 2020.

Yes, perhaps I'm off on yet another Man from La Mancha mission. No doubt, the devil will be in the details, transition, timing and economics. A million plans are being suggested, mostly in general conflict. Let's simplify the whole process by selecting this simple, but ultimate solution. Problem #3, which is that fatal flaw of our human society -- we can't seem to make grand decisions until it is too late -- can, thus, be partially overcome.

A critical mass of us making a stand today, with each just taking one constructive step, can galvanize our so-called leaders. *HuffPo* and the internet at large could be the key, for this new opportunity of instant feedback and propagation is changing the nature of decision-making. Marching the streets was so last generation. Impossible dreams have a way of now and then attaining reality. Maybe this one will save Humanity and Planet Earth.

Comments (16): There was a good discussion at the end. However, there remained some skeptical about overuse of energy if made free and a few couldn't quite comprehend that this energy would not actually be free because your taxes would pay for what you use. All in all, for such a ridiculous suggestion that we make renewable energy free, the response was encouraging.

On 29July08 I poked some fun at the extraordinarily large numbers being tossed around. Of course, our economic figures are nothing compared to the vastness of space. But Peak Oil and Global Warming remediation numbers are frightening, nevertheless. The International Energy Agency last month reported that it will take $45 trillion to insure that our climate only rises about 5 degrees Fahrenheit by 2050.[36] The Manhattan, Marshall and Apollo efforts combined, in today's dollar, only amounted to $0.24 trillion.

Billions and Trillions[37]

Carl Sagan wrote a book entitled **Billions and Billions**. While we tend to get wrapped up in our own infinitesimal problems, keep in mind that it takes light 100,000 years (travelling at 186,282 miles/second) just to get from one end of our Milky Way Galaxy to the other end. The closest spiral galaxy, Andromeda, is 2.5 million light years away. Yes, if we can ever design a spacecraft to travel at the speed of light, it would take two and a half *million* years to reach our closest true galaxy. And, there are more than 100 billion galaxies in our universe. But let's think small.

Modern man, *homo sapiens sapiens*, only arrived on the scene around 100,000 years ago. In a few years, the World population will approach 7 billion people. We first reached 1 billion just about 200 years ago, zooming to 6 billion in 1999.

Apparently, Everett Dirksen never did say "A billion here, a billion there, and pretty soon you're talking real money." Or, at least the Dirksen Center could not find anything in their files to verify that statement. Laid end to end, a billion one dollar bills would circle the globe at the equator four times. A trillion would take us from Earth to our Sun. One thousand billion dollars amount to a trillion dollars.

How far does a sum of one billion dollars go these days? We are paying $1billion/year to Pakistan for counterterrorism. There is a billion dollar large floating golf ball (really a radar station stationed in Alaska) that spends many holidays in Pearl Harbor. Each space shuttle shot is supposedly about a billion. The Lighthouse, a 984-foot skyscraper designed by American architect Thom Mayne, being erected in Paris, will cost a billion. The Freedom Tower (1,776 ft) has been financed for $3 billion. Each Nimitz class aircraft carrier carries a $4.5 billion tag.

What about rate of usage? For the past decade, the U.S. Department of Energy has annually spent less than a billion dollars for renewable energy research. The Bush budget for this coming year remains less than $1 billion. Farm subsidies will amount this year to $25 billion. And farmers are doing well because of the ethanol debacle. Our arms shipment to Middle East countries (not including Iraq and Afghanistan) will this year add up to $63 billion. We will spend around $600

---

36 http://www.environmentalleader.com/2008/06/07/iea-45-trillion-needed-to-cut-co2-emissions-50-by-2050/

37 http://www.huffingtonpost.com/patrick-takahashi/billions-and-trillions_b_115491.html

billion this year in the U.S. on gasoline. Nobel Laureate Joseph Stiglitz (Harvard economist) estimates the cost of the Iraq War to be $3 trillion, that is 3,000 billion dollars.

A few months ago, President George Bush requested a $3 trillion Federal budget, predicting a budget shortfall of $407 billion. But, it was this week reported that this deficit will actually be closer to $500 billion. Our overall Federal liability is $57.3 trillion, or about $500,000/household. Our gross national product is about $13 trillion, while that of the world is $65.6 trillion (we are 20% of the world).

How are our companies doing? Just four oil companies (Exxon Mobil, Shell, Chevron and ConocoPhillips) last year earned $100 billion in profits. Quarterly reports show that record profits are continuing this year. In 2007, General Motors lost a record $39 billion, and Ford just posted a worst ever quarterly loss of $8.67 billion. Toyota seems poised to sell more cars than GM this year. Our airline companies are predicting a total industry loss of $13 billion this year. But we're talking paltry billions here.

Going back in time, the Manhattan Project (to build a couple of atomic bombs) cost $2 billion, or $21 billion in current dollars. The Marshall Plan for post-war Europe cost us $13 billion over four years, or $80 billion today. The Apollo mission cost about $23 billion spread over 13 years, or $140 billion today.

So if Peak Oil and Global Warming are so dreadful, why don't we just start a new Manhattan Project for a sustainable world? Well, the International Energy Agency last month reported that it will take $45 *trillion* to insure that our climate only rises about 5 degrees Fahrenheit by 2050. The Manhattan, Marshall and Apollo efforts combined, in today's dollar, only amounted to $0.24 trillion.

Do we have a problem? Try dividing 0.24 into 45. You will obtain a figure of nearly 200, which means that for the world to take immediate action just to suffer an uncomfortable temperature rise would cost about 200 times more than Manhattan/Marshall/Apollo *combined*! Remember, the U.S. Congress this year killed the global warming mitigation bill and only included about a billion dollars for renewable energy research, while the G8 Nations in Japan weakened their resolve to tackle this problem. Oh, yes, our Congress, though, somehow, in no time at all, for them, actually found $300 billion to bail out our so-called housing crisis. What's going on? It really does not matter, for those were billions. The reality is that the $45 trillion IEA estimate could well be $450 trillion if you look closely at what actually has to be spent to control carbon dioxide by 2050.

Yes, the Pickens' $0.7 trillion wind farm initiative helps and so does the Gore $5 trillion renewable electricity for the U.S. proposal, which most pundits panned as impossible. Forget about billions and trillions. Don't waste time with more or less taxes. Why not just take that quantum leap to a *free* Green Energy Age, as explained in my ***HuffPo*** of July 17?

Comments (24): The responses were quite interesting and combative. A few agreed with me, but there were a few nit-pickers and some with valid counterpoints. I enjoyed the exchange.

On 2August2008 I returned to the subject of my initial HuffPo. As we search for funds within our budget, it is clear to me that the best place to raid is from national defense. Republicans, Blue States and the military industrial complex make this all but impossible. Yet, does anyone have any better ideas? Yes, cut taxes and become more efficient, but let's be real.

Why Do We Spend So Much On National Security?[38]

"I've never seen our lack of strategic depth be where it is today."
General Richard Cody, Army Vice Chief of Staff [39]

Let me see now, there is no USSR cold war threat. China spends $45/citizen for defense, while we invest about $2700/person on national security. Iran and North Korea are not global menaces. There are probably fewer than 100,000 terrorists, with a small fraction of them worthy of our concern. There will be no conquering enemy on the horizon for generations to come, if ever again. It was on this note that I submitted my first HuffPost on May 29, 2008 entitled, "Well, Barack, We have a Problem..."

How significant is national security in our Federal budget? Our fiscal 2008 discretionary funding is $941.4 billion. Defense and related accounts amount to $553.8 billion, but a supplemental sum of $306.6 billion needs to be added for our Global War on Terror and related needs. Thus, this year, we will spend $859.9 billion on WAR, much more than double what the Federal Government will expend on everything else! The Department of Energy will get $23.9 billion, of which about a $1 billion will be for renewable energy development, and the Environmental Protection Agency will spend $7.5 billion.

Is General Cody, maybe, exaggerating the truth? Actually, probably no, but not for a reason you might expect. With defense taking up so much of the national budget, you would think that we should be well covered to both defend ourselves and manage a ragtag bunch of terrorists. Well, our troop strength in the Middle East is below 200,000. Divided by our population of 304 million, this gives a ratio of 0.0006. In 1945, we had 16 million mobilized with a population of 140 million. The ratio then was 0.1143. In other words, if you divide .0006 into .1143, this would mean that we should be able to increase our total troop strength in this world hot spot by a factor of close to 200.

That comparison is almost meaningless, of course, for we have three million in uniform and reserve. But this makes you wonder what the concern is with only 6% of our available military actually in the Middle East, having had a period longer than World War II to make strategic adjustments. On an equal ratio basis with 1945, we should be able to mobilize 35 million, and GlobalFirePower.com points out that about 109 million are fit for military service in our

---

38 http://www.huffingtonpost.com/patrick-takahashi/why-do-we-spend-so-much-o_b_116535.html

39 TIME, April 14, 2008

country. Now that would really jack up the defense budget. Sure, this would mean a serious draft, but there is something about national service that deserves to be considered, anyway, for both genders.

All these numbers and analyses are interesting, maybe, but the whole point is, why are we spending so much money on national security? Is there a better way to gain the peace? We can talk about the military-industrial complex and their hammerlock over the White House and Congress. That's formidable, make no mistake about that. But perhaps the nature of world politics is such that the time has again come for us to mind our own business and invest in our national infrastructure and personnel. Maybe also do something about Peak Oil and Global Warming, too. Our presidential candidates talk about change, and our defense budget is a good place to start, providing the financial resources to actually do some real good. My initial *HuffPost* on "Well, Barack, We have a Problem..." provides a vision for this scenario.

Comments (9): I quote from my summarizing comment—

Interesting that there seems to be some consensus that we are, indeed, spending a lot, if not too much, on national security. There were no impassionate protests nor cries of anti-patriotism. Ergo... the funds to rebuild our infrastructure and combat Peak Oil / Global Warming can, perhaps, be conveniently drawn from a drastically reduced national defense budget. After all, WE HAVE NO THREATENING ENEMY ON THE HORIZON. The paranoia about China seems to yet dominate, so, as they have close to four and a half times more people than us, and spend $45/capita on defense, let's be sure and not match them in total defense expenditures, but double theirs on a per capita basis, meaning an American annual investment $400/capita. We can thus reduce our national security budget by $2300/person or around $700 billion/year (which, curiously enough, is close to the same amount we are expected to send to oil producing nations this coming year), most of which can thus be applied to all the needs just mentioned. If you missed my HuffPost of May 29, I actually boldly predicted how this might occur. And, of all the luck, Barack Obama returns home to Hawaii in a few days. Well, let's not get too carried away. I'll wait until he first becomes our POTUS.

*On 11August10 I decided to link a topic from **Book 1**, global warming, with one from **Book 2**, religion. Part of this composition is to further understand my mystification of why Americans are so prone to believe without proof. I find it difficult to comprehend why 90% of our citizens have faith in a heaven with no confirmed evidence, while less than 50% don't think much of global warming, when most of the responsible scientists say that this is real.*

Global Warming and the Afterlife[40]

Chapter 4 in **SIMPLE SOLUTIONS for Planet Earth** reports on global warming, and points out the greater peril of **THE VENUS SYNDROME**, with methane, not carbon dioxide, as the threat. Chapter 5 of **SIMPLE SOLUTIONS for Humanity** is a different take on religion, and wonders why 90% of Americans believe in the afterlife, while this figure is below 25% in most of Europe, Israel and Japan. The following synthesizes those two chapters, and anyone desiring details can go to those books, for 731 full references are provided.

Decision-makers are influenced by their constituency. If the public does not care that much about a problem, these leaders tend to ignore the significance of the issue. Conversely, if an overwhelming percentage of, say, voters, believe in something, then potential candidates take heed.

For example, most polls taken over the past few decades seem to indicate that around 90% of Americans believe in some afterlife and a God. It would thus behoove political candidates to be religious. The odds are, of course, that most of them actually believe themselves. At the national level, only Congressman Pete Stark is on record as not believing in God. However, he does not plan to run for office again. The fact of the matter is that something on the order of 90% or more of Americans would consider a black person or woman, but less than half would vote for an atheist.

As an interesting sidebar, only 5% of biological scientists in the National Academy of Sciences believe in the afterlife. Is there something this elite group knows that we don't?

Well, no. There are certain things that might not ever be known, while there are some other things that should be scientifically provable. Let's take global warming as an example. Ninety-seven percent of climate scientists believe that the global average temperatures have increased over the past 50 years, and 84% say that the cause is human-induced. Only 13% believe that there is relatively little danger to Planet Earth and us. Only 1% of them believe that the TV/cable media are very reliable and 3% rate local newspapers as very reliable. You can make your own conclusions on what these numbers mean.

---

40 http://www.huffingtonpost.com/patrick-takahashi/global-warming-and-the-af_b_117385.html

We all know that polls can be skewed by how you ask the question and who you solicit. So take the following any way you wish, but most recent surveys show that the American public is split about the environment being given priority over the economy and vice versa. In 2000, 67% favored the environment. Today, this percentage has slipped to 49%.

Last year 56% thought that cars and industry at large are mostly to blame, but the fault dropped to 54% this year. Slightly more feel that government should fine or tax company emissions, but only 52% to 45%. We tend to be concerned about water and air pollution, but only 37% worry about global warming.

Yet, Al Gore and others like him might be having an effect, as in 1997 only 25% said that global warming would pose a threat to their way of life (with 69% saying no), and this year the apprehension shifted to 40% yes and 58% no. All in all, though, climate change is not a huge concern to our populace.

I might add that a World Public Opinion poll reported this year that 43% of Americans felt that carbon dioxide was a pressing problem, while the returns from the world showed: Australia at 69%, Argentina 63%, Israel 54%, China 42%, Russia 32% and India 19%. Interestingly enough, 71% of those in the United Kingdom believed that this was all a natural occurrence and not a result of this gas. The recent UN Intergovernmental Panel on Climate Change report by more than 2,500 scientists found a 90% chance that people were the main cause and drastic action was needed to cut greenhouse gas emissions.

How then is global warming related to the afterlife? No, nothing to do with afterlife being hell, although that might make for an entertaining movie. Simply, for both, logic seems almost irrelevant. We tend to believe what we want, heavily influenced by our upbringing. Science is not welcomed in religion, for can you imagine the fate of a political candidate who might foolishly state: the greatest immorality of religion is that there is no proof of an afterlife. It's particularly worrisome that science is failing to have much effect on public opinion. There is no simple solution, but I'm trying, with Chapter 3 on education and Chapter 5 on religion in **SIMPLE SOLUTIONS for Humanity**.

Comments (11): I again quote from my summarizing response to one of the inputs—

I would agree with much, if not all, of what you said. The fact that 90% of Americans believe in a God and the Afterlife, probably helps to keep most people honest and caring, with the availability of an almost necessary psychological crutch for many. There is much value there. However, my purpose of linking Global Warming and the Afterlife was to stimulate conversation on how religion might, then, best take an active role in ameliorating climate change. I took on this self-proclaimed mission to Save Planet Earth and Humanity from Peak Oil and Global Warming by first writing those two SIMPLE SOLUTIONS books pictured on the right, then, reaching out into the ether through The Huffington Post. Common sense was not working, so I posted a two-part VENUS SYNDROME (Chapter 5 of SIMPLE SOLUTIONS for Planet Earth) nightmare, hoping that fear would work. That did not make a ripple. So I tried an out of box political solution by suggesting in a five part Huffington Post series to make renewable energy FREE. That never gained any traction. Thus, I thought I'd play the spiritual card. All you who are faithful, pass this post on to your congregation! If this fails, still to come is the joker (sort of the antipode to religion) in the deck: THREE STRIKES AND YOU'RE DEAD. Details on this concept can be found in Chapter 1 of SIMPLE SOLUTIONS for Humanity.

The presidential election was only a few weeks away, and while our unemployment rate on 25September08 was still 6.1%, it was obvious that something terrible was beginning to happen to our economy. Some say oil was a key reason for World War II, and that $147/barrel spike in July frightened the investment community. Many blame housing for triggering the crash, but like the previous serious recession in the early 1980's, behind this all was the metastability of petroleum.

Whoops, We Might Be On the Verge of Economic Calamity[41]

"The newly elected President blamed excesses of big business for causing the unstable bubble-like economy. Democrats believed the problem was that business had too much power and regulation of the economy was necessary." Is this Barack Obama in 2009? Nope, Franklin D. Roosevelt in 1933. The Great Depression did face an unemployment rate of 25%, which is now 6.1%. There was also an accompanying Great Drought beginning in the early 1930's. But the stock market largely recovered only a few months after October 29, 1929. Ironically enough, while the New Deal should be given some credit, the beginning of World War II in the later 1930's was the spark that boosted the economy, and there are respected historians who today partially blamed conflict over oil as an important catalyst for war.

Today, we are faced with the same, but more complex world, again spurred by petroleum. Crude oil futures experienced an all time high jump of $25/barrel on Monday, and settled at $121/bbl for the day. This was more than double the previous high spike of June 6, when the increase was $10.75/bbl. Okay, pundits blame the arcane future's trading process, not necessarily the Congressional mortgage bailout. Further, we can expect Congress to reign in that market, as it is doing with Wall Street. Sure, oil prices descended to $106/bbl the next day, but the metastable nature of the beast is ominous. Do you get a sense that we are beginning to lose control?

How does the $700 billion to $1 trillion bailout compare to other wartime expenditures and rescues? On July 29 I published a **HuffPo** article entitled, Billions and Trillions.[42] I reported that the sum total of the Manhattan Project (Atomic Bombs for Hiroshima and Nagasaki), Marshall Plan (to save West Berlin and post-war Europe) and Apollo Project (Man on the Moon) in terms of 2008 dollars, was roughly $240 billion. The amount of money actually spent was only $38 billion, but inflation relative those times brings this value in the range of a quarter trillion today, about a third the amount being discussed to resuscitate Wall Street.

First, where is this money coming from? Well, the combination of your tax dollars and foreign investments. If that was so easy, why didn't we solve our energy problem when voters were asking for relief a couple of months ago?

---

41 http://www.huffingtonpost.com/patrick-takahashi/whoops-we-might-be-on-the_b_128435.html

42 http://www.huffingtonpost.com/patrick-takahashi/billions-and-trillions-re_b_134702.html

Through the summer we all decried the sorry state of our energy condition and wondered why we were not more prepared, for we should have learned our lesson after the First Energy Crisis of 1973, but did not, and certainly after the Second Energy Crisis of 1979, but, again, did not. As we talk about billions and trillions, and reflect on the nearly $2 billion we were spending daily on gasoline in America in July and continued to look away at how much our federal government was allowing for renewable energy research (only an average of $1 billion/ year for the past decade), our Democratic Congress still could not even bother to renew those renewable energy tax credits nor approve any global warming rescue package. As early as June 14, my HuffPo was entitled, Piffle Squared,[43] lamenting the idiocy of this all.

Now, what makes the most serious environmental challenge ever presented to humanity and $147/barrel oil so insignificant? Why is saving Wall Street so important? Is it because the White House likes high oil prices and mocks the Greenhouse Effect? Could it be that Republicans can identify with high finance and were put into office by those who are today being rescued? Secretary of the Treasury Henry Paulson came from being the CEO of Goldman Sachs (which was also saved). Incidentally, he worked for President Richard Nixon, and more specifically, was an assistant to John Ehrlichman.

Nah, my sense is that we are in serious trouble and the White House just had to take those bold steps. I congratulate our leaders for taking action. I do, though, wonder about the attitude of the American Public when it comes to the looming dual hammer of Peak Oil and Global Warming. There is no National Will to do anything on this front. Somehow, I think that these virtual news portals might well be the solution to galvanize instant public reaction. Protest marches are so last generation. The world-wide web is the mechanism best suited to prevent what appears to be an imminent economic and environmental calamity. But how? You, out there, *help*!

Comments (8): Yes, most agreed with me that a mini-doom was to befall, and there was nothing we could do about it.

---

43   http://www.huffingtonpost.com/patrick-takahashi/piffle-squared_b_107137.html

Four days later, 29September08, the stock market did, in fact, crash, the Dow Jones Industrial Average dropping 778.[44] So my simple solution was for Congress and the White House to do nothing, giving good reasons why.

Simple Solution for Our Fiscal Mess[45]

In my two books I suggested SIMPLE SOLUTIONS for virtually everything: energy, environment, crime, war...name it, I had it. Except for the economy! Today I rectify that omission.

You might ask, what credentials do I have to even comment on the subject? Well, not much, actually, but more than I had in crime, religion and assorted other subjects which made the publications cut list. Plus, at least I once worked for the U.S. Senate, something most pundits just criticize.

Anyway, my **simple solution for the bailout is to do nothing**. This is in keeping with my samurai philosophy: act swiftly and decisively when you must, but sometimes, no decision is best.

Here is what our House members were thinking: only 40% of Republican voters, 30% of Democrats and 10% of independents are for the bailout plan proposed by President Bush and Treasury Secretary Paulson. In an Associated Press poll, only 30% of Americans said that they supported the Bush package. Yes, adjustments were made in Congress over the weekend, with a sprinkling of clever main street facades, but what is a Congressperson to do: vote for Wall Street and get kicked out of office on November 4? House Republicans, especially, kept to their principles (free market, not socialism...plus get re-elected) and the legislation was defeated 228 to 206.

Did you see what was happening to the stock market a minute before the debacle and for the rest of the day? The Dow Jones Industrials had already dropped more than 200 points, then suddenly, in a matter of a minute or two, plummeted another 500 points to minus 734 when it looked like the vote was hopeless, then, within minutes after the voting was supposedly closed (see Lawrence Berra's quote later about it not being over), when there came a reassurance that the House would re-consider and try to twist arms over the next couple of days over the Rosh Hashanah Jewish Holiday (first of ten repentance observances, culminating with Yom Kippur on October 9 this year), a gradual rise back to the minus 300s. The DJI then dropped again, largely in the minus five hundred range, and was at -580 seconds before the trading deadline today, when through a breathtaking few minutes after the bell (yes, Yogi's "It ain't over till it's over" wisdom prevailed again), the market free fell to 10,365, minus 778 for the

---

44   http://money.cnn.com/2008/09/29/markets/markets_newyork/index.htm?cnn=yes

45   http://www.huffingtonpost.com/patrick-takahashi/simple-solution-for-our-f_b_130433.html

day, the greatest one day drop in history. A decline of about 200 points AFTER the bell. My HuffPost of last week on the metastable state of our economy[46] is coming all too true.

So what will happen next? The White House will propose another package, the House will act by the weekend, a month before the General Election, and the Senate, at worst, might need to hang in there until next week, although the current plans are for the Senate to now deal first.. Is all this dilly-dallying all that bad for getting re-elected? Actually, no, for most campaigns are run on TV and sound bites. Incumbents can be made to look good with their home news channels reporting on the great job they are doing to save the country for their constituents. Plus, they won't need to pay for these self-promoting ads. What a windfall. Looks like our elected national representatives have found another workable ploy. The longer they stay in D.C., the more they will have to gain.

So returning to the DO NOTHING solution, let me ponder over a hypothetical nightmare. The following day, both the U.S. and world stock markets essentially went back to business as usual. Let's say our Congressional members cannot get their act together and leave town about mid-October, all the while blaming President Bush (who does not need to run this year) and the other party. There is enough to go around. Many banks close, but the feds take over, and most of the bank run is covered by the FDIC, anyway. Only the really rich lose bucks, but they caused all this, so no harm there. Gold could surge past $1000 / troy ounce, but no big deal here, because that also happened on January 21, 1980 (the Second Energy Crisis), and, in 2008 dollars, that's worth almost $2500/oz. Crude oil drops to $75/barrel because demand declines. OPEC countries squabble among themselves to reduce production. Then the price plateaus at $50/bbl, just enough to discourage the financial sector to cancel most large renewable projects. OPEC will love this. You can count of crude oil rising to more than $200/bbl in a couple of years. Paulson's Goldman Sachs predicted this. Then what?

As the stock market, the housing market slowly recovers. No $700 billion bailout package, but our existing financial system does have a certain resilience and flexibility to continue to provide business loans and assist consumer purchases of cars, homes and whatever. What was the emergency, again?

No way John McCain can get elected in this scenario, but Barack Obama will have at least a tiny mess on his hands come January 20. On the other hand, change comes best in time of crisis, and like FDR had Herbert Hoover, Obama will be able to thank George Bush. Doing nothing now might well be the optimal solution. The killing of the Emergency Economic Stabilization Act of 2008 was an early signal that the internet can be the controlling strategy for action. The virtual masses actually said no, and Congress shockingly agreed.

Comments (0): There were no responses. Of course, no one listened to me, and the U.S. Congress on October 1 passed the $700 billion Wall Street Rescue Package.[47]

---

46   http://www.huffingtonpost.com/patrick-takahashi/whoops-we-might-be-on-the_b_128435.html

47   http://www.consumersunion.org/pub/core_financial_services/006207.html

On 6October08 the Dow Jones Industrial Average at one point in the day crashed 800 points, again, a historic high, but settled at minus 370. Thus, I had to comment, so my HuffPo of 7October08 expressed some sarcasm to those decision-makers not listening to me. I blamed eight year of Republican rule, for their policy is to keep their hands off business. Well, that was at the root of this great recession.

The Bailout Bill Passes Congress, President Signs the Legislation, and the World Is Saved. Not.[48]

Well, on Monday October 6th, that crucial first Wall Street operational day after the rescue, the Dow Jones Industrial Average dropped a tad more than 800 points at one point--once again the *worst* one day drop in history--but settled at only minus 370, at 9956, about 30% lower than a year ago. For about the tenth time this past month in my daily blog I again underscored the danger of metastability. The fatal fear is that we are losing control of prices.

Reassuring, though, that the German, Paris and Tokyo stock exchanges have also all sunk 31% this past year, with London at minus 27% and China (Shanghai) a staggering minus 67%. They, too, all are bailing out banks and other financial institutions. Thus, you can't really only blame Wall Street, the White House, Congress, McCain and Obama. This seems to be a world-wide phenomenon.

Interestingly, precious metals have also dropped in the 30-70% range the past few months, even gold. Usually, investors switch to very low interest federal bonds or gold in time of coming crisis. Gold did rise a few percent today, but remains below what it was this past Friday. So there is some uncertainty in this reactionary front complicating the tea leaf readings.

The U.S., of course, sets the tone for the globe, and eight years of Republican free market policies no doubt laid the table for the crunch. The Bush Supreme Court aided by preventing state regulators from mortgage loan oversight (April 2007 decision). From all perspectives, this "keep the hands of government off business attitude" is the root of our financial crisis.

Ominously, my HuffPo prediction of $75/barrel oil seems to be happening, as NYMEX crude futures slipped below $90/barrel to $88.51/bbl, while the Brent Spot is now at $83/bbl. The growing world recession is reducing consumption, so there is weakening competition for the available oil. OPEC President Khelil today predicted that oil prices will continue to drop next year. Will we see $50/bbl oil in 2009? If so, don't be surprised if all those already announced billion dollar renewable resource projects begin to lose financial support. We saw it when Reagan became president in the early '80s and a decade ago when the price of oil (in real dollars) hit an all-time low. Here we go, back to the past again. See Chapter 1 *in SIMPLE SOLUTIONS for Planet Earth*. That's why it's ominous.

---

48   http://www.huffingtonpost.com/patrick-takahashi/the-bailout-bill-passes-c_b_132466.html

<u>Comments (4)</u>: *Again, I quote from one of my responses—*

I'm gratified that I have helped influence the thinking of at least one person out there. Only 6 billion or so more to go. I was surprised at the presidential debate today for Obama to actually state that energy was his #1 priority, especially with crude oil back down below $90/barrel. Maybe there is hope! Did you see those instant reactions to nuclear energy? Every time McCain touted this option, both men and women showed instant negative reactions. There seems to be a strong undercurrent of concern about this option. Likewise, when both mentioned renewable energy or doing something about global warming, the curve jerked upwards. Thus, the support is there. Unfortunately, the will is lacking. That has been the lament of all my HuffPosts. The simple solution, however, I still think, rests with those newfangled virtual portals...like The Huffington Post.

When I first heard Barack Obama speak at the 2004 Democratic Convention, even though he was yet a mere Illinois state senator, I was impressed. Learning that he was born and largely grew up in Honolulu sealed the deal. On 9October08 I felt compelled to point out what I thought was the only way he could lose to John McCain. Actually, my first draft was to call him platinum, to cast a glossy sheen to his potential. But I settled on the color mix, gray.

Barack Obama is Gray[49]

My very first HuffPost a few months ago was entitled "Well, Barack, We have a Problem." It was a wishful paean in search of the individual perhaps best suited to save our planet from global warming, while galavanizing world peace. In a Ray Bradbury *Sound of Thunder* reversal, I have come full-circle back to Barack Obama, but to a harsh real world on the verge of something worse than a recession.

Towards the end of CNN's post-presidential debate discussion on October 7, David Gergen declared it was too early to proclaim Barack Obama the victor of the 2008 presidential race simply because Obama was black. Polls are not totally believable, said Gergen, and Obama's blackness may cost him as much as six points. Gergen will be criticized, no doubt, for dealing this racial card, but he injected a very crucial point. This might well be the only factor left standing in the way of an Obama presidency.

Obama had his Reverend Wright, Khalid Al-Mansour and Tony Rezko. John McCain has been linked to the Keating Five, is on his second marriage and was operated on for melanoma in 2000. Nothing much is left to uncover. Yes, there is the upcoming third debate, but no game-changing surprises are anticipated.

The numbers are such that Obama should win. Yet, as that venerable Yankee catcher Lawrence Peter Berra might have said, "It will never happen until it happens." A non-white person has never been elected president of the United States. Pundits like to point out the Bradley and Wilder Effects, when black candidates for governor lost even though polls showed them ahead at the end. Gergen knows all this, and when I observed him making that fatal statement, his facial and body language seemed to be that of a positively concerned observer who very carefully felt compelled to blurt out this almost verboten fact. In my mind he did Obama a great favor. He hammered home the first nail on McCain's campaign coffin.

By all common sense, the people of the Nation should mostly vote for Obama on November 4:

1. Obama is 47; McCain is 72. Because of his bout with cancer, there are reports available hinting that McCain had only a 65% chance of surviving into the year 2010.

---

49   http://www.huffingtonpost.com/patrick-takahashi/barack-obama-is-gray_b_133101.html

2. Obama graduated #1 in his Harvard Law class; McCain was 894th out of 899 at the Naval Academy. If you were rating heart surgeons to operate on you, which medical equivalent would you choose?

3. Joseph Biden has a Juris Doctorate from Syracuse University; Sarah Palin meandered through five colleges over a six-year period, to finally graduate from the University of Idaho in journalism.

4. McCain supports President Bush in Iraq and the economy; Obama is for change.

5. Obama won the first two presidential debates over McCain.

The list can go on and on, but 66% of Americans are white and 13% black. All things being equal, people tend to vote their ethnicity. Hawaii has only minorities, but Filipinos vote for Filipino candidates and Japanese for Japanese. A person running for office actually gains when of mixed race, for a Chinese-Hawaiian will get most of the Chinese and Hawaiian votes.

This is where Barack Obama should have a huge advantage, for he is *both* black and white: his father is a PhD Kenyan, and his white mother was born in the heartlands of Kansas. Because she was busy saving the world in the Pacific and gaining her PhD at the University of Hawaii, Barack was in large part reared by her two white parents, and in Hawaii, where, again, there are only minorities. We are not a perfect society here, but equality trumps prejudice in our mélange melting pot.

There is a simple solution for the Obama campaign. Neutralize the Bradley/Wilder effect from the decision-making equation. Just make sure that the American populace knows that Barack Obama is Gray, or, better yet, both black and white, tinged with a variety of other colors from his upbringing and experience.

Comments (4): There were only 4 responses, but I thought I'd share mine--

I was chided that Obama did not really graduate #1 in his Harvard Law class. They changed the ranking system in the 70's and editor in chief of the Harvard Law Review does not now automatically mean that person heads the class. Sorry. However, what a feat to even be admitted to Harvard Law, then, becoming an editor, and finally, editor-in-chief, or number #1 editor. The part about McCain being #894 out of #899 at Annapolis, apparently, is perfectly accurate. Incidentally, I am disappointed that, while I got all kinds of personal comments from my reading audience, the only ones that made this posting had to do with the Keating Five. Is that what interests HuffPo readers? If so, that is, indeed, sad.

On 14October08 I revisited large numbers. Maybe if I say it often enough and **HuffPo** *keeps publishing it, someone will get the message.*

Billions and Trillions Revisited[50]

On July 29 I published an article in **The Huffington Post** entitled, "Billions and Trillions."[51] Let us today re-visit this subject in light of all the mega dollars being tossed around by pundits.

If gasoline costs $2.57/gallon (we used 142 billion gallons in 2007), we spend $1 billion a *day* on this fuel. Up to $4/gallon, the daily cost is a bit more than $1.5 billion/day. The current average national gas price is $3.16/gallon. Of course, crude is now back down to the $80/barrel range (or $1.90/gallon). Note the profit margin from what you are paying for gasoline. In Hawaii we are closer to $4/gallon.

The average annual U.S. Department of Energy renewable energy budget over the past decade has been less than $1 billion a *year* (Remember, we now spend $500 billion/year on gasoline). In comparison:

a. We pass on to Pakistan $1 billion/year for counter-terrorism activities, and they are not even helping much with finding Osama bin Laden.

b. Each Space Shuttle flight costs $1 billion.

c. Each new Nimitz class nuclear-powered aircraft carrier costs $4.5 billion. We will have 10 when the H.W. Bush is commissioned early next year. We have no naval threat into the foreseeable future.

d. Is it true that farm subsidies this year will be $25 billion? And farmers are now doing really well. Does the Farm Lobby spend $80 million/year on lobbying? In these good times for them and bad times for energy, why don't we shift just half of this $25 billion sum into renewable energy R&D?

e. The U.S. military is planning to move from Okinawa to Guam by 2014 at a cost of $15 billion. Why not just sell everything we can to Japan and send the troops to Afghanistan? By the way, it is reported that our military personnel have been involved in more than 200,000 accidents and crimes in Japan.

f. The top four oil companies made more than $100 billion in profits in 2007. General Motors lost $39 billion. Toyota sold more cars in the first half of 2008.

---

50 http://www.huffingtonpost.com/patrick-takahashi/billions-and-trillions-re_b_134702.html

51 http://www.huffingtonpost.com/patrick-takahashi/billions-and-trillions_b_115491.html

g. The Emergency Economic Stabilization Act of 2008 will cost $850 billion, or $0.85 trillion. That calculates out to be $2783/person.

h. Nobel Laureate Joseph Stiglitz estimates the true cost of the middle east war to be $3 trillion, six times more than reported by the Department of Defense. Over a seven year period, that would be about $1 billion/day. Did mention that over the past decade the Department of Energy has spent just $1 billion/year on renewable energy?

i. The International Energy Agency reported that it will take $45 trillion dollars to cut emissions by half to prevent global warming.

Are the above numbers too large to comprehend? Well, if you add the cost of the Manhattan Project, Marshall Plan and Apollo Project, then bring them up to actual 2008 dollars, the *combined cost* was $266 billion, or $0.266 trillion. (*You've seen this figure at $0.24 trillion because it depends on the year and exactly which economic index is being used*.) This total is less than a third of the Wall Street bailout package.

Of course, we're comparing apples and oranges because these atomic bomb / Europe saving / moon project funds were actually spent, while the fiscal rescue budget is sort of a loan. It is possible that we will gain a return over time for our personal $2783 investment. Sure.

Then, too, $0.266 trillion is about one half of one percent the sum needed to remediate global climate change. You will need 180 Manhattan/Marshall/Apollo equivalents to meet that challenge. What are we doing? Well, our Congress refused to pass the carbon cap and trade legislation and the G8 Nations merely begged Saudi Arabia to produce more oil. What about the general public? They seem content now that gasoline prices might soon drop below $3/ gallon. What a world! What priorities.

Peter Finch screamed in the movie *Network*, "I'm as mad as hell, and I'm not going to take this anymore." Any chance you *HuffPo* readers can help get something going?

Comments (2): *No one screamed. The only response was cut taxes, so I said—*

Great, cut taxes and preserve the free market. Doesn't that sound like eight more years of George Bush? Peak Oil and Global Warming are beyond the functional capability of the private sector. I'm afraid we need the G8 and United Nations working closely with industry to overcome what is looming as an even greater economic catastrophe for humanity. This latest financial mess will pall by comparison. Scarily enough, all this could occur very soon. At least you beg to differ. The masses don't care or don't know enough to be concerned.

Why I chose General Elections Day, 4November08, for this piece I don't remember. I guess this was a reaction to a report from MIT that there was more methane in our atmosphere. This simplest of hydrocarbons could seal our doom.

The Venus Syndrome Revisited[52]

On June 9, I published "The Venus Syndrome"[53] in the **Huffington Post**. In short, I expressed a fear that carbon dioxide might not necessarily be our greatest terror. I said there is something about methane.

Well, this past week, MIT[54] reported that methane has suddenly and uniformly increased in our atmosphere after a decade of general decline. The mystery was that this change was explainable for the northern hemisphere, the melting of the Artic tundra, of course. But what happened in the south? The usual suspects are mentioned, all terrestrially-based. That is a standard reaction of knowledgeable scientists. They tend to forget that the ocean covers about 70% of the planet Earth.

Did you know, for example, that there is more combined mass in the bacteria, viruses and archaea in the ocean than in all the larger life forms (fish, trees, you) in the ocean and on land? You've probably never even ever heard of archaea. Hint: like dark matter in space, archaea in our seas was relatively recently discovered, and virtually equals the mass total of bacteria.

The natural thing is that marine microorganisms at the surface expire, drop to the bottom of the ocean, and in an absence of oxygen, are converted into methane and other compounds, which, because of the pressure and temperature at depth, generally become trapped in ice as marine methane hydrates. It is said that there might be twice the energy in this methane at the seabed than all the known coal, oil and natural gas. Let me repeat: Twice as much energy in methane in metastable equilibrium at the bottom of the ocean than all the known coal, oil an natural gas deposits, which are rather safely resting deep underground. What happens to gas and ice when disturbed? Well, they rise to the surface.

Over our geologic history, every few ten million years, our planet naturally heats up. This is accompanied by heightened carbon dioxide and methane levels, or more probably, these gases caused the temperature rise... just like what is seeming to ensue today. Some scientists have speculated that the primary cause might well have been a rather sudden release of marine methane hydrates into the atmosphere. Chapter 5 of **Simple Solutions for Planet Earth** provides all the science and speculations an interested reader might want.

---

52   http://www.huffingtonpost.com/patrick-takahashi/the-venus-syndrome-revisi_b_140182.html

53   http://www.huffingtonpost.com/patrick-takahashi/the-venus-syndrome-part-o_b_106120.html

54   http://web.mit.edu/newsoffice/2008/methane-tt1029.html

Returning to that MIT report, is this methane increase a recent phenomenon? Actually, no. Since 1750 or so, carbon dioxide in our atmosphere increased by about a third, mostly, if not all, from burning fossil fuels. However, in this period, methane doubled! So there is something about methane.

But what's the big deal about methane in the air? Well, one molecule of methane is from 20 to 60 times worse than one molecule of carbon dioxide in causing global warming. According to some atmospheric scientists, this miniscule amount of methane in our atmosphere already has half the potency of carbon dioxide in warming our globe.

As an extreme worst-case scenario, let's take the case of our planetary neighbor. Venus is mostly carbon dioxide at a surface temperature of almost 900 degrees Fahrenheit. Could you imagine a scenario where sufficient methane contaminates our atmosphere to really cause trouble? Methane tends to oxidize into carbon dioxide over time. Ergo... the potential for an atmosphere and surface temperature on Earth like that on Venus.

Will this happen? I can't imagine that occurring, for our planet has been around for 4.5 billion years and we never got close to anything remotely that hot. Yet, could humanity's rush for progress stimulate a heating that, in combination with a perfect storm of events, catalyze the Venus Syndrome? Such a set of circumstances is presented in "Part Two of The Venus Syndrome"[55] from the June 10 Green issue of the ***Huffington Post***.

Comments (2): There was only one sarcastic remark. I guess there is no interest in the end of humanity on Planet Earth through a worst case scenario for global warming.

---

55   http://www.huffingtonpost.com/patrick-takahashi/the-venus-syndrome-part-t_b_106325.html

Republicans believe in world free trade. I tend to lean in this direction. However, there comes a time when you also need to look out for yourself. On 10November08 when I wrote this article, I thought the world was poised for a real depression. Yes, Buying American might well have triggered a world-wide economic crisis of confidence, but I floated this simple solution to gain a public reaction.

Buy American, Again[56]

The Democrats will be in full control over the next few years. There will probably be 58 Democratic senators when the dust clears, and two or three Republicans can be convinced to avoid a filibuster, if the issue is important enough.

The economy will dominate the decision-making for many months to come. A second economic stimulus package will be enacted, this year or early next. The American car industry, especially General Motors, is on the ropes. GM essentially declared in their third quarter report that it will become bankrupt by the middle of next year... unless there is Federal assistance.

Motor Trend has picked a car of the year since 1949. Until 1971 every winner was American. However, over the past six years, four have been Japanese, with only four of the seventeen cars currently making the finalist list for 2009 being American. They have already selected the Subaru Forester as the top 2009 SUV. Domestic maintenance records are actually improving, but we have time and again built the wrong cars with bad vision. This has to change. Recently passed was a $25 billion low interest loan to Detroit to work with the Feds in developing better next-generation vehicles. Let's do it right this next time.

Our economy has hopefully pushed through the equivalent of 1929. In 1933, the U.S. Congress passed the Buy American Act, which required the United States government to, yes, buy American. There were necessary loopholes and waivers, but this is one policy we might again consider. Republicans defend free trade, but Democrats, influenced by unions, have tended to support this sort of legislation.

Thus, the times and situation provide a simple solution to Buy American, recognizing as my two books did, that the implications are fraught with uncertainties. Even though Barack Obama campaigned on a platform of international cooperation, an explanation can be couched that America is the largest market for the world, and it is absolutely necessary that we, to avoid a total crash, for the good of the global economy, start by strengthening the purchasing foundation at home. To appease the Republicans, a bunch of trade barriers can actually be removed, on the premise that the populace will turn towards patriotism to Buy American.

---

56   http://www.huffingtonpost.com/patrick-takahashi/buy-american-again_b_142255.html

It might cost a bit more in the supermarket, the car might not run as smoothly, but there are good domestic cheeses and great American wines. In addition, we would be sensitive to the carbon footprint by buying locally.

Here, then, is the simple solution to our economic woes:

1. When the second economic stimulus package is passed, the President should boldly suggest that, as an act of loyalty to country, those planning to use this gift as down payment for a car, Buy American.

2. The U.S. Congress should attempt to pass the Buy American Act of 2009. Even if this fails, the whole point is symbolic. The buying public will need to be internally motivated to want to help and be part of the solution. An actual law might be counterproductive to international relations and difficult to enforce. I traded in my foreign car and bought a Ford when I went to work for the U.S. Senate. If I can do this, so can many others.

3. Barack Obama said throughout his campaign that we will need to sacrifice, initiate a mandatory public service opportunity for the youth and govern for change. Well, Buy American crystallizes the essence of his message into effective policy. Yes, we can, and will!

Comments (19): There were a lot of good comments, some agreeing with me, many not. One of my responses was—

I've noted my tendency in these HuffPo articles of mostly reacting to insults and the like. Well, Let me just say that newzzzjunkie is right. The whole point of my posting on BUY AMERICAN, AGAIN, is to let the PEOPLE take a stand and decide. Through most of my 25 or so articles in the Huffington Post I have continued to decry the lack of public will as the primary cause of our problems. Wake up Ameerica! We need all of us to each make that crucial difference.

Okay, so I thought for my next simple solution, let me link, Buy American, Green Energy and the incoming president, Barack Obama, posted on 17November08. It helps to connect a concept to what people have on their mind, and **Wall-E** *was just released, becoming the highest rated movie of 2008.*[57]

Wall-E, Eve and Barack[58]

It's far too early to predict how much change President-elect Barack Obama will bring into his new administration, but spotty floats certainly point toward same old-same old. But this is almost understandable so as to gain broad support and insure for a rational transition. Thus, the timing is ideal to suggest more visionary concepts as counterpoints to the seeming necessity of business as usual.

My very first HuffPo, entitled, "Well, Barack, We have a Problem," drew 15 mostly supportive comments. Someday, the challenge of Global Warming and Peace on Earth could well become the greater legacy, but for this current state of urgency, I will focus on the combined problems at hand: the economy, the automobile industry and energy, introduced in my previous HuffPo, "Buy American, Again."

The Wall Street rescue package is now known as the troubled assets relief program (TARP), which is further morphing into things unexpected. Detroit wants some of the $700 billion to help the Big Three, and recently, Phoenix, Philadelphia and Atlanta, representing cities, made a case to share in this largesse, arguing that public welfare makes the most sense, and mayors were at the ground level to immediately put people to work. Next, of course, the Terminator, weighing in to coordinate leadership among individual states, as California, in particular, is in terrible shape, only a step or two away from bankruptcy. It makes no sense to continue this strategy of charity.

So what's the simple solution? Certain magnificent obsessions can best become reality if the U.S. Senate convenes in 2009 with 60 senators, meaning no filibuster. With a looming visionary as the incoming POTUS (President of the United States), let us fantasize a bit.

In an inspired move supported by all, including the general populace, President Obama and our new Congress agree on the Economic Vitalization Edification (EVE) Act of 2009, endowing $2500 to each person living in the country (yes, maybe also illegal immigrants, but not tourists), with a stipulation that, if they purchase anything, it must be made in America. In addition, the legislation creates a Green Wall Street instrument, for now, called Wall-E, with the E, of course, representing a clean environment.

57 http://movies.toptenreviews.com/list_2008.htm

58 http://www.huffingtonpost.com/patrick-takahashi/wall-e-eve-and-barack_b_144190.html

Wall-E would have a portfolio of incentives only for environmentally enhansive companies. Detroit's Big 3, for example, would need to spin off corporate entities (and this might well be *one* joint firm) focused on renewable energy powered vehicles. Yes, this all sounds awfully close to that lovable Waste Allocation Load Lifter Earth-class, WALL-E, and the love of his life, EVE (Extra-terrestrial Vegetation Evaluator), so maybe get Steve Jobs and Disney somehow involved in the marketing of the concept.

Simple Solutions for Planet Earth, published last year, predicted that oil would drop to $50/barrel, and, unfortunately, all things sustainable will again be shelved. So, what happens? The Brent Spot Price last week made it down to $51/bbl. On cue, T. Boone Pickens has apparently, for now, abandoned his wind farm efforts, supposedly because credit is tight, but, let's face it, $50 oil significantly raises the risk factor, the deadly virus for renewable energy projects. There is little hope for $100/bbl oil until some time later next year or henceforth, but, while wise decision-makers this time will not be trapped into thinking all is now well, the critical timing factor is two months, for whatever the Obama transition team prioritizes could well sustain his administration for the full first term.

So, the next economic rescue package must transcend current unreality and focus on saving these Green programs because we all know that when the world financial crisis is overcome, oil prices will zoom past $100/bbl, perhaps even up to $200 in five years. If these sustainable energy projects die now, it will take many years for all of them to recover, way too late to meet the next sure thing energy crisis, again. Advocated is a new paradigm for this challenge, where each of us patriotically buys American and invests in Planet Earth, a true partnership of WALL-E, EVE, Barack and us.

Comments (2): Not much input, but my entry said—

BWILDER gets it! Now, how can we get a critical mass of HuffPo readers to similarly take just one more step by sending this message on to your mailing list? Changes can best come from crises, so we are in luck, for you are at an unprecedented point of your life. How many of you were alive in 1929 (hint: you need to be almost 90)? Today, however, we can compound this recession heading towards depression with the dual hammer of Peak Oil and Global Warming. This temporary drop of petroleum prices, of course, can only depress interest in renewable energy investments, plus, those snow flurries I see on TV unfortunately make too many yearn for warmer weather. So, OPEC and Mother Nature are joining forces to test you all. Through all of my HuffPosts I have lamented the lack of public will in making a true difference. Somebody out there, help find a way to overcome complacency, economic depression, climate change and soon to skyrocket fossil fuel consumption by introducing President-Elect Obama to WALL-E and EVE. Now!

On 21November08 HuffPo published my effort at promoting a long-shot ocean energy option. I don't think much about wave or current power, but I've long harbored hopes for ocean thermal energy conversion, for the Blue Revolution can only develop if this renewable energy source is commercialized.

The Coming of OTEC[59]

There is a renewable energy option that might have been first dreamt 138 years ago by Jules Verne in his **Twenty Thousand Leagues Under the Sea**. Ocean thermal energy conversion (OTEC)[60] features sustainable baseload power -- meaning it is continuously available, as differentiated from solar and wind energy, which are intermittent -- and is so vast that tapping less than one-tenth of one percent of this stored energy in the ocean (which would be continuously replenished if utilized) would supply more than 4 times the total amount of electricity consumed in the World. Initially proposed in 1881 by Jacques d'Arsonval, a French engineer, his student Georges Claude (who also invented the neon tube, but was later sentenced to life in prison for being be a Nazi collaborator) failed in the 1930's off Cuba to prove the concept. Finally, Lockheed, off Keahole Point on the Big Island of Hawaii, first produced net positive power in 1979.

I just happened to be working for U.S. Senator Spark Matsunaga in D.C. that day, so helped draft the first OTEC bill, which was almost immediately passed by the U.S. Congress and signed into law by President Jimmy Carter. We might have been overly optimistic, but the legislation predicted that 10,000 MW of OTEC power would be in operation by 1999. For the record, this total is today zero.

In 1981, a Japanese consortium led by Tokyo Electric Power Company succeeded in feeding to the Nauru grid 120 kW of closed cycle OTEC power. Toshiba produced two 10 minute clips, Part I which can be viewed through You Tube.[61] You can then click on Part 2. Unfortunately, a hurricane wiped out the facility. I have annual dinners with the group to re-live these incidents, and there were many.

Well, I returned to Hawaii in 1982 and helped invent the Pacific International Center for High Technology Research (PICHTR), suggested by Senator Matsunaga and named by then Governor of Hawaii George Ariyoshia. My engineering team at PICHTR, through funding from the U.S. Department of Energy and Japan, succeeded with a 255 kw (gross) open cycle OTEC facility[62] at the Natural Energy Laboratory of Hawaii on the Big Island, also providing freshwater.

---

59 http://www.huffingtonpost.com/patrick-takahashi/the-coming-of-otec_b_145634.html

60 http://www.nrel.gov/otec/what.html

61 http://www.youtube.com/watch?v=_mGOcqofERM

62 http://www.otecnews.org/articles/nelha_otec_history.html

India, aided by Saga University of Japan, made a 1 MW attempt in 2006,[63] but had cold water pipe problems. Otherwise, most of the notoriety with the technology has been associated with mere announcements of projects on Diego Garcia (Department of Defense), various Pacific Islands (Department of Interior and Japan) and the Caribbean (Solar Sea Power), never attaining fruition. A current summary of the growing field can be found in OTEC News.[64]

Now, however, Hawaii has again gained the spotlight with an announcement that Lockheed Martin, with the Industrial Technology Research Institute (ITRI) of Taiwan, is designing and will build a 10 MW OTEC facility, most probably to feed electricity and freshwater to Honolulu. Doug Carlson has an OTEC blog[65] providing details. How things come back full circle, for it was 20 years ago that Paul Yuen, PICHTR president, and my engineering team worked with ITRI on an $80 million multiple product OTEC facility proposal.

Perhaps now the Blue Revolution (Chapter 4 of *SIMPLE SOLUTIONS for Planet Earth*) has finally begun, with promise for renewable energy, green materials, exciting habitats, marine biomass plantations, next generation fisheries and, maybe, remediation of global warming and prevention of hurricane formation.

*Comments (0): I guess this was too arcane and esoteric for **HuffPo** readers, I guess, as there were no comments.*

---

63 http://www.ioes.saga-u.ac.jp/english/about-india-otec_e.html

64 http://www.otecnews.org/

65 http://hawaiienergyoptions.blogspot.com/

If there is any concept I'm most involved with today, it is the Blue Revolution. Nothing much will happen for me to witness, but within the century, I see the potential for what should be the next frontier for Humanity, the open ocean. Thus, on 24November08 I posted:

The Dawn of the Blue Revolution[66]

This is Part II of my series on ocean resources, which was initiated with my previous *HuffPost* on "The Coming of OTEC" announcing the development of a 10 MW OTEC facility by Lockheed Martin for Honolulu.

Many of you have heard of the Green Revolution, which improved farming practices for grains. The Blue Revolution, while similar in concept, is much more than just a marine copy, for there is promising potential for producing a wide range of sustainable products -- renewable energy, green materials, exciting habitats, marine biomass plantations, next generation fisheries -- while actually improving the ocean environment, through, perhaps, remediating global climate warming and preventing the formation of hurricanes.

Twenty latitude degrees above and below the equator is an ocean region currently considered to be a wet desert, for nothing much grows near the surface. This is because of a lack of nutrients. However, at 1000 meter depths is deep ocean water at 4 degrees Celsius with nitrogen and phosphorous concentrations, for example, respectively, at 100 times and 20 times greater than what is generally found at the surface. This is because life in the photic zone eventually settles at the ocean depths and decomposes in the exact ratio as is required for life at the surface. The deeper ocean is cold because of natural convection from the Arctic and Antarctica. Interestingly enough, marine biomass, certainly in the cellular form, is said to be from two to five times more efficient (in converting sunlight to mass) than any land crop (trees, grasses). Part of the reason is that land plants need to draw nutrients through a root system, while marine life can utilize the entire surface area contacting the ocean.

Natural upwelling brings some of this rich fluid to the surface over only one tenth of one percent of the ocean, where nearly half the seafood is harvested. If a means can be found to bring some of this deep ocean water to the surface, ocean ranches and farms can be supported, which would not need to be fed (the growth cycle would be closed), fertilized or irrigated. Another advantage is that the open ocean is generally considered to be free. The timing is ideal because all fisheries are now reported to be declining, at a time when demand is increasing because of population growth and changing nutritional trends. This resource also offers hope for the production of liquid biofuels and hydrogen. Part 3 of this series will delve into Next Generation Fisheries.

---

66   http://www.huffingtonpost.com/patrick-takahashi/the-dawn-of-the-blue-revo_b_145889.html

It so turns out that the temperature differential between 1000 meter and surface waters can be utilized to bring huge quantities of water to the surface in a manner which sounds suspiciously like a perpetual motion machine. Ocean Thermal Energy Conversion (OTEC -- see previous posting) has been developed to the stage where net positive electricity can be produced. Hawaii happens to be the site where most of this development has occurred, and the Natural Energy Laboratory of Hawaii Authority is home to more than 30 companies and research organizations supported by deep ocean water. Japan already has a dozen such facilities.

It is the total package of products, though, including seafood, biopharmaceuticals, green materials, biofuels, etc., that appears to have niche applications of commercial interest. In the very long term, there is reason to believe that each grazing OTEC plantship could form a city or, even, country. Chapter 4 **of *Simple Solutions for Planet Earth*** provides details.

In my 2003 Bruun Memorial Lecture to UNESCO[67] in Paris, I proposed that the United Nations take a leading role in galvanizing Project Blue Revolution. There are important Law of the Sea and international political issues to be considered. There are today only 192 countries forming the UN. Someday, perhaps, a thousand OTEC-powered Blue Revolution platforms, each a nation in itself, could well be plying our oceans, providing clean and sustainable resources for Humanity in harmony with the ocean environment.

Comments (2): There was only one input saying we should replenish fish stocks by banning fishing. I responded thusly—

I've attended numerous hearings and international sessions where 90% of the participants decried the current practices as causing the declining conditions. Of course they are right. However, while I can understand why you are saying what you advocate, in principle, I abhor fines, limits and anything that limits freedoms. The whole point of the Blue Revolution and Next Generation Fisheries (see Part 3 to be published later this week in this publication) is to INCREASE production so that we can have enough for everyone...within limits, of course. Please comment on my Part 3 when it appears.

---

67   http://unesdoc.unesco.org/images/0013/001352/135278e.pdf

We are moving along with renewable electricity. We are slow to develop sustainable ground transport. We are doing almost nothing about aviation, but that is a future topic. For now, my 28November08 posting touched on all.

Simple Solutions for Our Biofuel Problem[68]

Renewable energy will be used to provide electricity, and fuels for ground transportation and aviation. Sustainable electricity is well in hand. As the price of oil again increases, the range of alternatives will gain in prominence, starting with windpower, which is already competitive. Plug in vehicles are poised to make an impressive entry.

Liquid and gaseous biofuels are another matter. Details about this subject can be found in **SIMPLE SOLUTIONS for Planet Earth**. The current infrastructure is liquid dominated, so, for this discussion, I will delete biomethane and hydrogen from consideration. In addition, for now, hydrogen is just too expensive to produce.

The Nation and World unfortunately went in the wrong direction when ethanol from corn (plus sugars and other starches) and biodiesel from terrestrial plants were selected for focus. The Farm Lobby no doubt should be congratulated for smartly lobbying Congress and the White House, for farmers are in great shape. Job well done! Regrettably, grain prices jumped, causing a food crisis for developing countries. The knee-jerk reaction was, of course, all that fibrous cellulose, why not ferment those wastes into more ethanol? So, a second herd of white elephants is now being groomed. Why? Because there is a simpler alcohol called methanol that makes more economic sense. My HuffPo article of June 10 compares ethanol and methanol.

Regarding biodiesel, the notion is almost laughable, as only a very small percent of the plant itself is used. Biodiesel from algae, for example, is ten to twenty times more efficient in converting sunlight into usable fuel. Plus, these land plants grow relatively slowly and need irrigation water.

About the future of cars, one always hopes for nanotechnology making a difference, but it appears that the lithium battery might well be the ultimate. Per unit volume, fuel cells can provide more energy than any battery, perhaps by a factor five or more. This is why the direct methanol fuel cell (DMFC) will in time replace batteries for portable electronic applications. As an aside, this defies common sense, but one gallon of methanol has more accessible hydrogen than one gallon of liquid hydrogen. Thus the logic argues for producing methanol from biomass to power a fuel cell. Whoops, there is no DMFC for vehicles. For the record, while methanol has half the energy value of gasoline, the fuel cell is at least twice the efficiency of the internal combustion engine, so there is a wash here on fuel storage. And methanol is no more toxic than gasoline. You shouldn't drink either one.

---

68 http://www.huffingtonpost.com/patrick-takahashi/simple-solutions-for-our_b_146906.html

Aviation remains a great challenge. I wrote into the original Senate legislation, known as the Matsunaga Act,[69] a section that did result in the hydrogen powered National Aerospace Plane Project in the 1980's, which later became a black Department of Defense (DOD) Program. Thus, the only option over the next few decades has to be a replacement for jetfuel.[70] The Defense Advanced Research Projects Agency has a solicitation for a commercial jetfuel from algae,[71] and companies are coming out of the woodwork to do the job. Is this good? Yes and no. Nice that the DOD thinks $3/gallon jetfuel is attainable, and terrific that industry is eager to comply, but the basic science and engineering has not yet been attempted. Why not?

First, the National Oceanic and Atmospheric Administration (NOAA) is mostly a monitor and protection agency. Second, the Department of Energy abandoned all ocean energy efforts and the National Renewable Energy Laboratory actually gave their microorganism collection to the University of Hawaii. The National Science Foundation did establish a Marine Bioproducts Engineering Center in Hawaii, but only for high value products. Energy is cheap, and that is the problem with biofuels from algae.

So what are the simple solutions to develop a progressive national biofuels program?

1. Terminate Federal support for ethanol from food.

2. Adjust the existing language for tax incentives to say, "ethanol, biodiesel **and other biofuels from renewable resources**." This would make available future funds and permit academics to conduct research on biomethanol, biobutanol, etc., and stimulate industry involvement in these areas.

3. Quickly perfect a direct methanol fuel cell for vehicular applications. A direct ethanol fuel cell will be inherently inefficient.

4. Expand the mission of the Department of Energy to bridge the gap between research and commercialization and permit R&D for areas that link to energy. The economic potential of many renewable energy areas is enhanced with non-energy co-products.

5. As NOAA reports to the Department of Commerce, expand the current research policy to encourage oceanic research to commercialize marine products and closely partner with the USDOE to share common interests.

6. Mobilize a national program to accelerate the development of ocean energy, paying special attention to advancing biofuels production from microorganisms.

---

69   http://virgobeta.lib.virginia.edu/catalog/u1771298

70   http://www.renewableenergyworld.com/rea/news/article/2008/11/the-quest-for-alternative-fuel-in-the-aviation-industry-takes-off-54069

71   http://www.biofuelsdigest.com/blog2/2008/03/06/darpa-says-ahead-of-schedule-to-produce-jet-fuel-from-biomass-at-less-than-3-per-gallon/

7. Accelerate the development of next generation aircraft, whether it be a hydrogen dirigible that can be engineered to travel 500 miles per hour or advancing the twenty year old National Aerospace Plane Project, or both.

The Obama energy transition team is now determining priorities for the next four years. With oil prices loitering around $50/barrel, I worry that attitudes have shifted about the seriousness of our future energy problems. We, of course, know with a high degree of certainty that oil will zoom past $100/barrel again, if not next year than certainly within five years. The development of sensible next generation biofuels will take a decade, so, in a sense, we are already too late. But the timing is perfect for change, so let's do it the right way this time.

Comments (14): The discussion was encouraging. Most agreed with me and I actually made a couple of important contacts. Anticipating the upcoming debate about the direct methanol fuel cell, I quote from one of my responses—

By the way, a couple of people mentioned to me that my article left the reader in midstream. Why, they asked, is there no such thing as a direct methanol fuel cell for vehicles. Very simply, as ethanol is the only sanctioned national biofuel, research funds are not provided for anything to do with methanol. Mind you, the convincing logic of the concept is such that INDUSTRY has already just about perfected the Direct Methanol Fuel Cell for portable applications. Toshiba has announced a product for sale to begin replacing batteries in computers and similar applications. Now, it's just possible that the DMFC for cars will never be commercialized because the challenges of finding the ideal membrane or otherwise significantly improving the efficiency will be too great. But the tragedy is that we just don't know and were prevented from making an effort. TO THE OBAMA ENERGY TRANSITION TEAM: you absolutely must give the DMFC and the renewable methanol economy a chance.

An early step in the Blue Revolution is development of next generation fisheries. In addition to aquaculture, marine versions are starting in the coastal region using cages and feeding with fishmeal. My 1December08 posting focused on "The Ultimate Ocean Ranch."

The Ultimate Ocean Ranch[72]

This is part three of a series on the Blue Revolution, today focusing on those repeating headlines indicating that there is a very serious decline in fish stocks[73] throughout our oceans. Projections show that even with the increasing world population and a shift of nutritional patterns away from red meat towards seafood, actual fish production will decline in the future. Already seafood costs more than chicken, pork and beef in supermarkets. At one time not too long ago, ocean products were a bargain. Remember those Friday fish meals (having something to do with some Christian belief) which were barely tolerated? The increasing gap between supply and demand will even further raise prices.

Aquaculture[74] already supplies half (70% from China) of marine food production. But the future is in the sea around us, and mariculture, meaning aquaculture in the open ocean, is increasing, albeit slowly, because of environmental and other constraints.

Nearly a decade ago, I was co-author of "The Ultimate Ocean Ranch,"[75] a concept to farm the ocean utilizing the following practices:

1. Place these ranches in the open ocean away from coastal environments.

2. Link to the cold water effluent of ocean thermal energy conversion (OTEC) plantships. These waters are very high in nutrients in the exact ratio and composition needed for sea life.

3. Conduct the basic science and engineering for the process:

   a. Utilize nutrient or temperature barriers (i.e., no cages).

   b. Close the growth cycle so that no feeding is necessary (see #2).

   c. Acoustically harvest the seafood (not too adventuresome, but sound can attract fish).

   d. Develop the robotics to protect the fish.

---

72   http://www.huffingtonpost.com/patrick-takahashi/the-ultimate-ocean-ranch_b_146192.html

73   http://www.wri.org/publication/content/8385

74   http://fisheries.ifcnr.com/article.cfm?NewsID=696

75   http://findarticles.com/p/articles/mi_qa5367/is_199908/ai_n21443379/

e.   And so on.

Next generation fisheries never did gain any traction, but, with the advent of OTEC, there is now promising hope for this concept.

I've certainly tried, for I co-chaired a National Science Foundation sponsored gathering in 1991 on Research Needs for Off-shore Mariculture Systems, and chaired the first international summit in Tokyo on the subject involving Japan, Norway and the U.S. in 2004. The following year I helped orchestrate the Bergen Declaration[76] on Next Generation Fisheries, which was also signed by Chile. Thus, much of the international understanding and scientific storyboard to advance the field are already in place.

The clear leader for Next Generation Fisheries is the Pacific International Center for High Technology Research,[77] headquartered in Honolulu, Hawaii. Anyone out there interested in joining the mission to better feed humanity and insure for the survival of fish species nearing extinction should contact them.

Comments (2): There was a favorable response from Marcel F. Williams, who has an excellent blog site,[78] and regularly comments on my articles. We might not agree on everything, but we do find interest in the same subjects.

---

76   http://www.rundecentre.no/docs/BergenDeclaration-NGF.pdf

77   http://www.pichtr.org/

78   http://www.newpapyrusmagazine.blogspot.com/

The price of oil will determine the next decade of world solvency. If petroleum drops to $20/barrel, this will kill all renewable energy projects, but time will then be provided to better develop these options until they can later be commercialized, my posting of 7December08. While in disbelief of Bloomberg's prognostication, and not in sarcasm, but with a positive approach, I posted the following on:

A Gift to Planet Earth and Humanity[79]

A **miracle** has occurred. Many were beginning to contemplate a survival strategy because of the dual hammer of Peak Oil and Global Warming. But a funny thing happened on our way to doomsday. It is appearing that we are getting a reprieve, and, ironically, the gift is this serious, but fixable, economic collapse.

An absolutely incredible prognostication in Bloomberg is that January delivery will see crude oil below $20/barrel[80] and oil traders are today purchasing gasoline for $0.97/gallon. What does all this mean?

Let us look at some historical consequences. Much of this is detailed in Chapter 1 of **SIMPLE SOLUTIONS for Planet Earth**, but the price of gasoline in 1973 (all the following will be in 2008 dollars) was $1.80/gal when the First Energy Crisis increased the price to $2.30/gal. The Second Energy Crisis of 1979 kicked it up to $3.17/gal in 1981. Gasoline then, with a small uptick due to the Gulf War of 1991, declined to **$1.35/gal in 1998..** (This is the equivalent of 37 cents/gallon in 1973.)

On July 11, 2008, petroleum skyrocketed to $147.27/bbl, causing gasoline to sell for $4.12/gallon on July 17. Gasoline on December 5 subsequently crashed to $1.77/gallon, 57% lower. The stock market only declined about half that in this interim.

This time, then, the economy affected oil prices, as the recession is reducing use, causing an oversupply. Following the First and Second Energy Crises, higher oil prices dampened the economy. After recovery, oil (and gasoline) prices dramatically dropped. The key question is, will this next recovery raise oil prices? Yes, of course.

We can thus expect the following:

1. Oil could drop below $30/bbl.

2. Gasoline prices might plunge below $1/gallon. **This will be the lowest gasoline will ever be in all of history, even if the decline only settles down to $1.34/gallon.**

---

79 http://www.huffingtonpost.com/patrick-takahashi/a-gift-to-planet-earth-an_b_148902.html

80 http://www.bloomberg.com/apps/news?pid=20602099&sid=aGVEWl4Nnlio&refer=energy

3. If no Solar Manhattan effort occurred in 1982 when gasoline prices were near the modern day high, what are the prospects of anything monumental happening when it will be at an ALL-TIME LOW when Barack Obama becomes President on January 20? Remember, T. Boone Pickens, with all his sincere bluster, abandoned his wind farms when oil was still more than $50/barrel.

4. The economy will recover. It might take all of a year if there is no depression, but, certainly, in 5 years.

5. If, in the meantime, oil does rise to $75/barrel in a year or two, the slack now available to producers will delay any sudden escalation, for OPEC can increase production as necessary. However, the surging economies of China, India and rest of the world will almost surely further escalate the price beyond $100/barrel, maybe up to $200/barrel, in five to ten years. There is such a thing as Peak Oil, and there is mounting evidence that the Middle East does not really have as much of this resource as they claim.

In grand summary, then, this world economic collapse could well be a gift to Planet Earth and Humanity, for Peak Oil will be delayed, carbon dioxide in our atmosphere will be somewhat alleviated and, thankfully, we will have this five to fifteen year period to work on sustainable options that can begin to competitively replace fossil fuels.

The following simple solutions can be recommended:

1. Take advantage of this "gift" of time and comprehensively prepare for a sustainable energy economy. The Obama energy transition team and the new Congress can either instill a yes we can change...or royally blow it.

2. As research and development are only a fraction of actual commercial investments, government can cost-effectively and should expeditiously partner with industry and academia to plan for, fund and implement a visionary renewable energy mandate, even more monumental than the Apollo Project.

3. Smartly insert a carbon tax linked to the price of oil, now. As crude prices increase, so will, thus, this tax. At $30/barrel, something like a 2 cents / pound carbon dioxide tax is significant but tolerable. When oil eventually jumps to $150/barrel, the tax should be proportionately higher, at 10 cents / pound carbon dioxide. The revenues should fund renewable energy and conservation programs.

4. To kick-off the Planet Earth rescue strategy, immediately add a $1/gallon gasoline investment surcharge (also known as a tax, but the semantics can't hurt--and remember, Europe and many parts of the world already pay twice as much for pumped gas), which will result in nearly $150 million/year, also to be applied to the Obama Sustainable Energy Plan.

Society barely reacted after the First Energy Crisis of 1973. We did absolutely nothing after the Second Energy Crisis in 1979. Let us use this "gift" to act wisely this time.

Comments (13): Again, I appreciated the various inputs. I quote from one of my responses—

I agree with all you said. The whole point of this posting, though, is, IF THE REAL PRICE OF OIL DROPS TO THE LOWEST LEVEL EVER IN HISTORY, as seems destined to occur next month, will the Obama Administration follow through on their good intentions? On June 2 I published in the Huffington Post Why is there No National Energy Policy? The problem identified was the will of the people. Issues such as global warming and Peak Oil have just not been important enough to most of us. I worry that the free fall of gasoline costs will again serve to delay any monumental effort to end our addiction to oil. Then, when oil shoots past $100/ barrel again, as it surely will, we will wonder what happened. Are we smart enough to do it right this time?

It turns out that oil only settled down to $35/barrel, and quickly more than doubled to the $75/ barrel desired by OPEC.

I've vacillated on the promise of electric cars. Yes, the hybrid Prius was a breakthrough, and I think a combination of battery and fuel cell replacing the internal combustion engine will be the mode of ground transport for the next century, or more. However, I don't think this will occur for several decades because hydrogen is just too expensive and lacking in infrastructure. But an all-battery car which you can plug in and charge using your home electricity troubles me because of a fatal flaw in this technology. I thus felt compelled to post the following article on 15December 10.

Is There An Option More Promising Than The Plug-In Electric Vehicle?[81]

Thomas Friedman recently published an opinion piece in the **New York Times** entitled, "While Detroit Slept,"[82] equating any congressional or presidential rescue of the Detroit auto industry to saving the mail-order-catalogue business on the eve of eBay or improving typewriters just before the advent of the personal computer and the internet. In his mind, the Big Three has been anachronistic, and entrusting them with an eleven-digit taxpayer loan would be foolish. He is probably right, even though the specter of a Depression triggered by their bankruptcy nevertheless cannot be totally discounted, so our domestic auto industry will no doubt be given one more chance.

He muses that the internal combustion engine/gasoline transport system is approaching obsolescence, and other concepts such as Shai Agassi's Better Place electric vehicle network model a more promising future. Appropriately enough, an agreement was announced early this month for Hawaii to be one of their first demonstration sites. A few days later, Maui Electric Company and Phoenix Motorcars signed a memorandum to use their electric pick-up trucks. All this is well and good, for Hawaii, naturally blessed with all the renewable energy options, has, for a variety of reasons, lagged behind much of the nation and world in going green.

A predictable trend is, no doubt, a gradual shift to battery-powered cars which can be charged with wind and solar energy. The lithium battery is poised to serve as this power source.

I say, let us support this effort, but be watchful for two impacting factors, one bad and the other, possibly monumentally good. First, the bad: my *HuffPo* article of December 7 on "A Gift for Planet Earth and Humanity" worries that the petroleum excursion below $50/barrel has, maybe fatally, dampened large-scale investments of renewable energy for a long time to come. T. Boone Pickens' abandonment of his wind farms is only one of many such crushed ventures you'll be reading about in the months to come. Crude oil, of course, will again shoot pass $100/bbl, if not next year, then certainly in five. However, financial organizations,

81   http://www.huffingtonpost.com/patrick-takahashi/is-there-an-option-more-p_b_150824.html

82   http://www.nytimes.com/2008/12/10/opinion/10friedman.html?_r=2

rightfully so, are allergic to risk of any kind, and fickle oil prices have historically bedeviled all solar options. Thus, read the above posting to take advantage of this "gift" of time so that a range of remediative strategies can be applied.

The second factor has to do with the long-term viability of battery systems. It is possible that lithium might well be the end of the line. So, in answer to Mr. Friedman, an earlier **HuffPost** on "Simple Solutions for Our Biofuel Problem"[83] suggests another next generation technology as maybe a more hopeful choice rather than plug-in vehicles.

Per unit volume, a fuel cell should be able to provide five times more energy than the lithium battery. Chapter 3 of *Simple Solutions for Planet Earth* provides the details on fuel cells, but, in short, this device works like a battery to produce electricity, but uses hydrogen as the energy source instead of lithium, lead or cadmium. However, and this defies common sense, **one gallon of methanol has more accessible hydrogen than one gallon of liquid hydrogen.** Thus, the logic argues for producing methanol from biomass to power a fuel cell, as hydrogen is very expensive to manufacture, store and deliver. This simplest of alcohols is the only biofuel capable of directly and efficiently being utilized by a fuel cell without passing through an expensive reformer.

Yes, methanol has only half the energy value of gasoline, **but the fuel cell has at least twice the efficiency of the internal combustion engine,** so there is a wash, here, regarding onboard storage. And methanol is no more toxic than gasoline. You shouldn't drink either one.

But we have problem. The U.S. Department of Energy has prohibited providing funds for vehicular direct methanol fuel cells (DMFCs), and furthermore, stopped supporting biomass to methanol R&D. It has mostly to do with ethanol and biodiesel being selected as the only national biofuels. Perhaps the Farm Lobby might have also had a role to play in this decision. Thus, we are probably a decade away, if not longer, and maybe never, from being able to convert to a biomethanol economy for transportation.

Thus, unless some sudden advancement can be realized in bringing a transport DMFC to the marketplace, it makes sense to cultivate options such as the plug-in electrical car system, hoping that electricity from the renewables can enjoy a quick commercial transition. In any event, watch out for the direct methanol fuel cell, for this virtually ignored opportunity could well either someday replace vehicles powered by batteries or in parallel maybe develop even faster.

Comments (32): There were a lot comments, and the most persistent was an individual who supported another fuel cell technology. He did not think that the direct methanol fuel cell would ever become real. However, it is already being marketed by Toshiba for portable applications. I did learn, though, that there was a lot more resistance to the plug-in electric car, as summarized in one of my responses:

---

83   http://www.huffingtonpost.com/patrick-takahashi/simple-solutions-for-our_b_146906.html

Wow, yours is just the kind of response I had on my priority wish list. I was sort of wanting to be nice and not be a wet blanket for plug-ins because my State and the national policy are totally supportive of them. Detroit, too, appears to be leaning in this direction. HOWEVER, IF PLUG-IN ELECTRICAL CARS ARE A DEAD END, then the new Obama sustainable energy plan should purposefully abandon electric cars and go immediately into the renewable hydrogen powered fuel cell vehicle. While there remain serious questions about how much this fuel cell will cost when commercialized, if competitive enough, then the question becomes, METHANOL or HYDROGEN as the fuel. My heart is for clean hydrogen, but my engineering sense argues for methanol, as the infrastructure is already largely in place, it should be cheaper to produce, and this unreal fact: one gallon of methanol has MORE HYDROGEN than one gallon of liquid hydrogen. Thus, the highest priorities should be to perfect a direct methanol fuel cell for vehicular applications and development of a more cost effective way to gasify and catalyze biomass into methanol.

In any case, my current sense is that the battery-fuel cell hybrid will catch on in a decade as the vehicle of choice. There remains a large question about what fuel or fuel cell. I remain convinced that the direct methanol fuel cell would be best.

*My first **HuffPo** was a recommendation to Democratic Presidential candidate Barack Obama. My 33rd 18December08 was to President Obama.*

A Solution for Barack[84]

Pardon me for being so informal, but, we're both from Hawaii, and this title follows my very *first **HuffPo*** on May 29, "Well, Barack, We Have A Problem..."[85] which boiled down those two **Simple Solutions** books into a whimsical musing about how Barack Obama could fund his plan to combat Peak Oil and Global Warming, and in so doing, end wars forever. Since that article, he went on to prevail over Hillary Clinton and John McCain, and is now the President-Elect, while I proceeded to post 32 more pieces for the **Huffington Post**, all on energy and the environment, and wondering why I was doing this.

I woke up at 3AM this morning and realized why. All the above crystallize the essence of what should be our national energy policy,[86] and I might well have a solution for our President-Elect. Don't blame President Bush, or the Congress, or the oil companies for our energy problem. We are at fault. You, me, us. This posting drew more than 100 comments, arriving to the conclusion that we have no energy policy because it is not important enough.

Oil is now below $50/barrel. Gasoline prices are approaching historic lows (in 2008 dollars). As sincere as P-E Obama was about green jobs and change, complacency looms to prevail. HuffPo #32 suggests that our current economic collapse actually is "A Gift to Planet Earth and Humanity."[87] After the Energy Crises of 1974 and 1979, there was a flurry of activity, followed by a general abandonment of anything sustainable, as we went back to our petroleum addiction. We now have the worry about the Greenhouse Effect, plus, that $147/bbl spike in mid July, so, hopefully, we can be smarter this time.

But the Obama energy transition team (OETT) has the daunting task of delivering a meaningful plan against the tide of low oil prices and competing priorities. And what a doozy, for one of my other HuffPos on "Billions and Trillions"[88] reports: if you combine the total cost of the Manhattan, Marshall and Apollo programs, and bring it up to 2008 dollars, you need nearly 200 times more money than what those three monumental efforts cost our nation, just to meet the challenge of the Greenhouse Effect.

But you've got to start somewhere, so while the OETT won't have trillions of dollars to spend, they can make a crucial difference. There will be a million ideas and thousands of lobbyists

---

84 http://www.huffingtonpost.com/patrick-takahashi/a-solution-for-barack_b_151822.html

85 http://www.huffingtonpost.com/patrick-takahashi/well-barack-we-have-a-pro_b_104201.html

86 http://www.huffingtonpost.com/patrick-takahashi/why-is-there-no-national_b_104507.html

87 http://www.huffingtonpost.com/patrick-takahashi/a-gift-to-planet-earth-an_b_148902.html

88 http://www.huffingtonpost.com/patrick-takahashi/billions-and-trillions_b_115491.html

clamoring for attention. More than anything else, the team needs to clarify and focus with vision. I offer the following simple solutions:

1. Renewable electricity is almost a given, but maintain those tax incentives and expand research on many of the solar and ocean options. Add a penny per pound carbon dioxide tax, proportionately linked to $30/barrel oil, to adjust as oil prices vacillate, so that at $150/barrel, the tax should be 5 cents.

2. Ground transportation is a real problem. **A dollar per gallon investment surcharge on gasoline can be justified.** We'd still be paying much less than in Europe and Japan. Ethanol from food, of course, should be terminated as fast as the Farm Lobby allows, no, make it much not sooner. The fermentation of cellulose to ethanol should be compared against gasification/catalysis into methanol. Unfortunately, a fair assessment might not be possible because the Department of Energy does not support any biomethanol research. Unfortunately, too, my HuffPo #33 and subsequent comments from this readership suggest that while the current darling, plug-in electric cars,[89] are the obvious bridge for the next decade, they might be a dead end. However, the fuel cell option will need hydrogen, which is very expensive to produce, handle and store, or the direct methanol fuel cell, which might have a fatal flaw. What then? Good luck! **But at least take a close look at the methanol economy**.

3. Aviation? This has not even been considered, but should. When I drafted the first hydrogen bill for Senator Spark Matsunaga nearly three decades ago,[90] I added a clause for the National Aerospace Plane, thanks to input from Lockheed. To shorten a long tale of delay, we are at least 25, if not 50 years, away from a next-generation hydrogen powered jetliner or cost-effective jet fuel from marine algae. Are we then in deep trouble? Hawaii especially, but I've learned of a new concept, for now, let's call it the **Hawaiian Hydrogen Clipper (or H2 Clipper)**, which the advocates say, can be developed in a decade. While the prudent might be skeptical, I support the idea because there is nothing else on the horizon, and this idyllic spot in the Pacific would, indeed, be the ideal location to pioneer this development, for we have the political clout -- P-E Barack Obama was born and grew up in this State and Senators Daniel Inouye (chairman of the full Appropriations Committee) and Senator Daniel Akaka (on Energy and Natural Resources Committee) represent Hawaii -- natural resources, and definite need, for without a next generation sustainable air travel alternative, when jet fuel prices again skyrocket, tourism will truly collapse, and economic depression will follow.

So my plea to the OETT is to press forth with vigor, and let me know if you would like to hear more about the H2 Clipper. Aloha.

---

89 http://www.huffingtonpost.com/patrick-takahashi/is-there-an-option-more-p_b_150824.html

90 http://www.huffingtonpost.com/patrick-takahashi/what-about-free-hydrogen_b_110532.html

Comments (10): *The responses were uniformly supportive. I even got praised, to which I responded:*

Thanks. Nice to be complimented. Now, if the rest of the world can be drawn in to read these postings and buy my books, that would be a fine way to begin to save Planet Earth and Humanity. More importantly, I really thought that the Huffington Post, with that instant response potential, could galvanize real action, and make obsolete protest marches. This has not worked, but I suspect I'm part of the problem because I have not made the effort, for example, to link with Blogger, Twitter, WordPress, TypePad and Tumbir...whatever they are. Well, this is a good a time as any to get started on those.

Aloha.

On Christmas Day of 2008 I sent to HuffPo my Holiday Message. Did you know, for example, that the two best selling songs of all time relate to Christmas?

My Holiday Message[91]

Since May 29 of this year when I wrote my first article for the Huffington Post, to my most recent on December 18, both with Barack in the title, I have focused on attempting to find Peace for Humanity[92] and a solution for Peak Oil and Global Warming.[93] On this, my 35th posting, my holiday message, while maintaining a somewhat provocative edge, provides a bit of humor, followed by compelling evidence on how the Catholic Church might have originated.

I will first draw from the world wide web ten nearly useless bits of holiday trivia in the spirit of my SIMPLE SOLUTIONS books found in one of the boxes on the right. It is possible that absolutely correct references or attributions might never be found for some of them. A few, not unlike miracles, will need to be taken on faith.

1. To complete his tasks, Santa needs to travel between 3 and 5 million miles per hour, depending on who you quote. This is theoretically possible because the speed of light is on the order of 670 million miles per hour.

2. There is a contention that Santa used female reindeers with male names because they (boy deers of this species) lose their antlers in the winter. Females don't...but yes, they do have them. The fact that neither gender can fly is irrelevant to this discussion.

3. Rudolph the Red Nose Reindeer was created for Montgomery Ward in 1939 as an advertisement gimmick. A decade later, Gene Autry sung the song that became the second best-selling of all time. White Christmas, by Bing Crosby, remains #1.

4. In the 12 Days of Christmas, 364 presents were given. The 2008 value can be averaged to be just under $20,000. This period runs from Christmas Day to January 6 (Epiphany, meaning "to manifest or show," when there is a Christian feast to celebrate God's "manifestation" in human form...Jesus.) Then, Christmas can again be celebrated on January 7 by switching to the Julian calendar.

5. Something in the range of half a million annually contract food poisoning eating spoiled Christmas meals and snacks.

---

91 http://www.huffingtonpost.com/patrick-takahashi/my-holiday-message_b_153502.html

92 http://www.huffingtonpost.com/patrick-takahashi/well-barack-we-have-a-pro_b_104201.html

93 http://www.huffingtonpost.com/patrick-takahashi/a-solution-for-barack_b_151822.html

6. If you did not give your dog a Christmas present, you were not only a Scrooge, but in the gross minority.

7. Jingle Bells was written in 1857 for Thanksgiving and was the first song broadcast from space, as performed by Gemini 6 astronauts, Tim Stafford and Wally Schirra, on December 16, 1965, with the full accompaniment of sleigh bells and a harmonica smuggled on board, while reporting on a command module they saw piloted by someone in a red suit pulled by eight smaller modules.

8. The Christmas tree was first popular in Germany as early as the 8th Century. O Tannenbaum (Christmas tree) came from 16th Century lyrics and an old German folk song. It takes anywhere from 4 to 15 years to grow a 6 foot tall fir tree.

9. Sir Isaac Newton (mathematician) and Little Richard (singer) were born on this day, while Billy Martin (baseball) and Dean Martin (singer) passed away on Christmas.

10. There is a nasty rumor that Santa has an evil brother, Satan Claus, who lives in the South Pole, and makes fruitcakes.

On a more serious note, excerpted from Chapter 5 of **SIMPLE SOLUTIONS for Humanity,** is a hypothetical tale reporting on, perhaps, the most significant mortal (or God) ever, Jesus Christ:

...place yourself back to circa 50 AD. Nothing much can be found about Jesus. Conflicting descriptions can be unearthed about this individual, but absent any miracles and the connection with Son of God. Is it possible that a group, say, the Antioch Jesus Movement, sees an opportunity to spur something called Christianity? So they pick a mortal of those days who might just fit a concept called the Messiah. They borrow selectively from early Egyptian, Indian, Chinese, Mexican (nah, scratch this, Columbus came 1500 years later) and Greek writings to create the legend of Jesus Christ around a real-life martyr. He was in his prime at the age of 30, so they choose 30 AD as the founding of Christianity. Could this incredible PR ploy have started it all? Did all this lead to the Catholic Church of today?

If the above sounds patently ridiculous, or, maybe, sacrilegious, much of what is written in **The Bible** is taken by many as Gospel, with equally flimsy documentation. In any case, to intelligently comment, you need to read the full chapter to appreciate the boundaries of the available evidence. Merry Christmas and a Happy New Year!

<u>Comments (2)</u>: *I guess few read or feel like commenting on Christmas Day.*

I'm trying to figure out why I even wrote this article. I guess I was reacting to an all day night electricity blackout in Honolulu, the fact that Black Friday refers to that shopping frenzy after Christmas, and thought I'd project out to the ultimate modern doom predicted by the Mayans.[94] *I wrote this the previous day, but it was published on 27December08.*

Black Friday[95]

I was just about to post the following when a lightning storm caused a Honolulu blackout, which lasted all of 20 hours for me. Friday must have also been black in terms of shopping, as I could not find parking space at the Ala Moana Shopping Center. Thankfully, the stock market did not suffer a Black Friday, but actually went up, which, I guess, can be considered to be a profitable or black day.

On another black matter, Michael Casey provided a relatively upbeat summary, if that's possible, of the 2004 Indonesian tsunami,[96] which killed 230,000 four years ago, triggered by the second largest earthquake ever measured. This was by far the most deadly catastrophe of its type (death from seismic waves) in all of human history. Apparently, both nature and society are recovering well so soon, considering the enormity of the tragedy.

The #1 earthquake of all time occurred in 1960 off Chile, with a moment magnitude of M 9.5. The above Indonesian monster was between M 9.1 and M 9.3. At one time, the Richter scale was used, and provides lower numbers than the now more popular moment magnitude. A magnitude of 9.0 is ten times more powerful than an 8.0. If these earthquakes occur in the ocean, then tsunamis can result. However, the Chilean version "only" killed less than 2,000, including 61 in Hawaii, where the wave crested at 10.7 meters (35 feet).

Can there be anything worse from the sea? Yes, for in 1970, half a million perished from a cyclone (same as hurricane or typhoon) devastating East Pakistan. This cataclysm was at least partially responsible for this region seceding from the country and becoming Bangladesh in January of 1972.

Can there be anything even worse from the ocean? Again, yes, but not in a way you might expect. Should there be a mega-earthquake (say anything larger than M9.0) in the bottom of the sea, a near shore wave can be as high as 100 meters, but only about a max of 10 meters in the far field (a thousand or more miles away). However, if there is a mega landslide of sufficient size and velocity falling into a deep ocean, the tsunami can be as high as 1000 meters, depending on who you ask.

---

94 http://december212012.com/articles/mayan/index.shtml

95 http://www.huffingtonpost.com/patrick-takahashi/black-friday_b_153737.html

96 http://www.huffingtonpost.com/2008/12/26/2004-tsunami-victims-reme_n_153573.html

Lituya Bay in Alaska 50 years ago experienced a 524 meter (1720 feet) wave, when Howard Ulrich and his son on board the Edrie were carried into the woods and survived. It was reported, though, that the wave was less than 75 feet when it struck the boat.

The most hyped possible event is the La Palma Mega Tidal Wave,[97] featured as a potential reality[98] by BBC News and regularly shown on the Discovery Channel. The result of Cumbre Vieja Volcano collapsing into the sea could produce a 900 meter (2950 feet) colossus, striking Florida, New York and Boston with 50 meter (164 feet) waves. Eminent scientists disagree on this projection.

Islands are most prone to this event because many form at the bottom of a deep ocean from volcanoes, which can naturally be eroded by water action. The island on which I live, Oahu in Hawaii, has supposedly suffered from more major landslides than anywhere else. The Nuuanu Landslide has been mentioned as one possibly having caused a mega tsunami a million or so years ago. For those who have been here, driving to the other side of the mountain chain from Waikiki, you will gawk up at the Koolaus, which represent the inside of a major crater. The rest of the volcano can be found in the ocean behind you. What is particularly disarming for me is that I live on Nuuanu Avenue.

As I was one of the geothermal reservoir engineers for the Hawaii Geothermal Project a third of a century ago, I had students build a model of the Big Island to determine how these steamy pockets form. Well, it then occurred to me that if certain rifts happen to join, say, triggered by a major earthquake, a good portion of that island could theoretically fall into the ocean, perhaps causing a mega tsunami. I picked a period four years from now, August 2012, for this doomsday event in Chapter 6 of **Simple Solutions for Planet Earth** and entitled it, "Six Hours to Seattle." I don't want to give away the ending, but I can reassure you not to lose any sleep, especially if you live in Hilo.

Chapter 5 of the same book, incidentally, selected August 12, 2012, which happens to be another Black Friday, as a different type of doom, the mere end of civilization through the Venus Syndrome. If we survive that day, then the 13-baktun cycle of the Maya reaches the termination point on December 21, 2012, which, then, would be a particularly Black Friday if the prophesy actually happens.

Well, four years from now is about as far away as that devastating Indonesian event in 2004, so we are talking real time. Of course, religion aside, how can any reasonably sensible person believe in these undocumented myths and scares?

Comments (4): Someone commented about the passing away of her Great Dane as why this was a Black Friday. We all have different reference points for what is doom.

---

97   http://www.lapalma-tsunami.com/tsunami.html

98   http://news.bbc.co.uk/2/hi/science/nature/1513342.stm

On this final day of 2008, December 31, I pontificated about the future, all based on my assorted **HuffPo's** *of the past year. Part of my concern was that the price of oil had fallen to $35/barrel, and that we were again entering the price crash era of the late '90's when the cost of petroleum dropped to an all time low.*

Simple Solutions for 2009[99]

By now you have no doubt read a dozen reviews of 2008 and projections for 2009, all pure guesses for the latter, unless someone was carefully predicting that the sun would rise tomorrow and the like. I would like to, instead, affect the future by sharing a few simple solutions. Here are my top five:

1. <u>End to the Gaza Strip War</u>: "Why don't those #&%#@* Hamas idiots just stop lobbing rockets into Israel," courtesy of a golf buddy pundit expressed at the Ala Wai Golf Course Clubhouse. He wanted me to carry his views in my daily blog, but I thought I would protect him from any possible fatwa by not specifically identifying the source. White House spokesperson Gordon Johndroe expressed those exact sentiments, with a safe political spin.

2. <u>To initiate the process for Peace on Earth, forever</u>: My very first *Huffington Post* article, circa May 29, 2008, entitled, "Well, Barack, We have a Problem...",[100] can be paraphrased to say: *...go to your very first G8 Nations Summit, by your declared emergency to be held at United Nations headquarters in New York City, and pronounce a Gorbachev-like bombshell: our country will reduce military spending by 10% this year, and will continue to do so for the next eight years. This scenario is described on page 65 of Simple Solutions for Humanity shown in one of the boxes on the right. You say, we want every country to do the same, for this 10% solution is our best opportunity to create a global fund to combat Peak Oil and Global Warming (PO/GW). At this stage, keep quiet about the "ending wars forever" part, as then, no one will take you too seriously. China's knee-jerk reaction might well be, what, cut defense spending? We haven't had a chance yet to attain your level of capability. But, on quick afterthought, they will realize that they will only need to decrease their spending by $6 billion in Year One, while the U.S. takes a $60 billion hit. Ten percent of the worldwide $1.2 trillion/year for war means that at least $120 billion/year will suddenly become available in the first year to overcome PO/GW.* This sum comes close to the annual amount suggested by the International Energy Agency to meet the total $45 trillion recommended by the International Energy Agency earlier this year to cut the carbon dioxide level in half.

3. <u>Saving the American auto industry</u>: Each of us in this country has already loaned GM and Chrysler $44 (for a total of $13.4 billion), which could well end up being almost 10 times

---

99   http://www.huffingtonpost.com/patrick-takahashi/simple-solutions-for-2009_b_154609.html

100 http://www.huffingtonpost.com/patrick-takahashi/well-barack-we-have-a-pro_b_104201.html

more if the situation deteriorates. The best that GM could offer about their future was a plug-in Volt electric car. Why copy the Japanese? Think, really think, about the future of ground transport, and develop a truly promising next generation vehicle. Reference can be made to my posting on "Is there an Option More Promising than the Plug-in Electric Vehicle?"[101] This alternative future is the direct methanol fuel cell. Unfortunately, there is no such widget today, for the U.S. Department of Energy currently bans funds to develop this technology, and also for any biomethanol production research. What better partnership, then, for the auto industry and farmers, with government incentives, to manufacture the renewable fuel and power-train system to lead the world in a decade?

4. <u>Saving the national economy</u>: There will be an Obama stimulus package, perhaps with a trillion dollar value. The details are provided in "Buy American, Again",[102] but the timing is right for our new president to help Main Street when, on President Bush's watch, $863.4 billion were provided to Wall Street, porkers and Detroit. An equivalent amount divided among our citizenry would then be just under $3,000. To save a few bucks for other needs, Congress and the White House should allow $6,000 to each couple filing a tax return and $3000 for individuals. However, don't send a check. Just assign each couple/individual a credit card for the appropriate amount, which can only be used to buy products made in the USA. There are some downsides to this plan, but they are explained away in the indicated article.

5. <u>Saving Planet Earth and Humanity</u>: About global warming, I've published 35 articles in the ***Huffington Post***, and all of them, to some degree, treat this subject. Let me then simplify with two specifics. First, immediately mandate a dollar a gallon investment surcharge on gasoline before oil prices rise again. Gas will still cost much less than European rates. Second, add a penny per pound carbon dioxide disenhancement charge (this will add 2 cents/kWh to your electricity bill if your power comes from a coal-fired facility), proportionately linked to $30/barrel oil, to adjust as oil prices vacillate, so that at $150/barrel, the tax should be 5 cents/pound. Yes, some of the revenues will need to be directly rebated to the people, but a good fraction should be utilized to develop sustainable resource options.

What are the odds of any one of the above attaining reality? Zero, unless each of you sends this posting to ten colleagues, who....

<u>Comments (6)</u>: *There were a few good ideas offered, but it is clear that President Obama is not listening to me. Sometimes I wonder why I even bother.*

---

101 http://www.huffingtonpost.com/patrick-takahashi/is-there-an-option-more-p_b_150824.html

102 http://www.huffingtonpost.com/patrick-takahashi/buy-american-again_b_142255.html

5January09 was still before the inauguration of Barack Obama. My frequent postings were primarily to influence the transition process for an administration inspired by the need to affect change. This was yet another attempt at catching the eye of a key staff member.

The Sun will Continue to Shine...[103]

...at least for another 5 billion years. The survival of humanity as we currently know it will, however, be tested. This is the feedback I received from a stream of collegues to my most recent article on Simple Solutions for 2009.[104] However, interestingly enough, they chose not to comment in *the Huffington Post*. At the end of this posting, I suggested that my article be shared with their contacts, and it is this "chain-letter" like strategy that stimulated dialogue. The thread of communications particularly focused, not on the economy, but on Peak Oil and Global Warming.

To summarize, a few felt that it is already too late, doomsday is destined to come, and maybe soon. A few more are reconciled to a lower standard of living. Some believed that monumental changes might still be possible, but we need inspired leadership and major structural adjustments to government. My latest response to one such input follows:

Dear XXXX:

You're, of course, absolutely right. In my second Huffington Post article on June 2, entitled, Why Is There No National Energy Policy?,[105] I went so far as to say that we shouldn't blame George Bush, nor the U.S. Congress, and not the oil companies. We have no energy policy because we don't feel it is important enough. The problem is us, as Pogo could have said. There were more than a hundred comments.

It's like education. We grumble and complain, but generally do nothing much about the poor scores of our students in those K-12 international test. Why? Because, as a nation, we are doing fine. We are the greatest nation ever, and the reason why is that we have by far the best higher education system, spending more money than any other country where it does make a difference. You need 5% of the populace to lead the way, but maintain an adequate core of followers, sort of what we are doing. So don't expect any sudden increase in salaries for lower education teachers because that's the smart thing to do for our future. We might, though, look closer at Finland and Singapore, and borrow as necessary.

Essentially, we need a serious crisis. Hitler and the fear that Germany would develop the A-Bomb first was the reason for the Manhattan Project. Sputnik resulted in the Apollo

---

103 http://www.huffingtonpost.com/patrick-takahashi/the-sun-will-continue-to_b_155160.html

104 http://www.huffingtonpost.com/patrick-takahashi/simple-solutions-for-2009_b_154609.html

105 http://www.huffingtonpost.com/patrick-takahashi/why-is-there-no-national_b_104507.html

Project, leading to the end of the Cold War. I thought we reached the energy equivalent with $147/barrel oil a few weeks after the above article. Alas, crude oil at less than $50/barrel will make it all but impossible for any drastic and accelerated change. Global Warming? Many millions will need to perish for any real movement.

So, I'll maintain my pontification crusade to continue the education process, awaiting the day when $200/barrel petroleum or that really hot summer arrives so that we are better prepared to make that crucial difference. In the meantime, I still think there is something to the direct methanol fuel cell and the biomethanol economy as a necessary bridge, plus the Hawaiian Hydrogen Clipper. For now, enjoy life as best as you can. The Doomsdayers might turn out to be right, but I'll continue to valiantly plug on hoping for the best.

Aloha.Pat

Is the inability of our leaders to act with decisiveness on a matter of global significance in the absence of a major crisis the fatal flaw of our current society? Are we destined to soon be crushed by the dual hammer of Peak Oil and Global Warming, crippling our civilization forever? Is there anything about the incoming Obama Administration that provides any confidence for Planet Earth and Humanity, especially as that other brushfire known as the economy will almost surely occupy most of their time and political capital? However, as time is of the essence, let's just hope that the trigger will not be a catastrophic crisis. The Sun will continue to shine, but will you be part of the solution?

Comments (10): *The reader who is anti-direct methanol fuel cell returned with a vengeance. He could be right, but I still suspect that his basic motivation was to undercut the competition to his type of fuel cell, which colleagues tell me has no future. Otherwise, no signs of life from the Obama transition team.*

Having read in the paper that yet another $4.5 billion aircraft carrier was to be christened, I wondered, why we were building so many of these big ticket defense dinosaurs when there is no one to fight anymore. That led to analyzing large numbers in general, following up on two previous postings on billions and trillion. Thus, why not quadrillions?

Trillions and Quadrillions[106]

My *HuffPo* of 29July08, entitled, Billions and Trillions,[107] provided an almost startling number: $240 billion or $0.24 trillion. That was the total amount the United States spent, in 2008 dollars, for the Manhattan Project (to bring a quick end to World War II), the Marshall Plan (which built the foundation for a strong Western Europe against communism) and the Apollo Project (which was the seed that bankrupted the Soviet Union, leading to the end of the Cold War). Startling, because that was cheap, recalling Congress just setting aside $700 billion to rescue Wall Street and Barack Obama negotiating a trillion dollar stimulus package with the Congress.

On October 14, I followed with Billions and Trillions Revisited.[108] I noted that the average U.S. Department of Energy annual budget for renewable energy research over the past decade was under $1 billion. The brand new nuclear-powered aircraft carrier, the USS H.W. Bush, which will be commissioned on 10January09, cost taxpayers $4.5 billion. This will be our tenth Nimitz-class carrier, and the last.

However, on the horizon are the Gerald R. Ford class nuclear carriers, the first to cost $8 billion, not including the $5 billion R&D being expended to insure for the producing the finest. It's a toss-up if the initial version will be called the Gerald R. Ford or America. The second has already been reserved as the USS Arizona. We're now up to $21 billion to scare the Russians and Chinese into they, likewise, investing in their future.

Do we have a vote on these matters? I have a problem with big ticket Defense expenditures in these days of rinky-dink terrorism. But the first one of this new series is still on the drawing board. Obama Team: you want to initiate change and reduce (or, at least delay) spending? Here is one for consideration.

But to go on, the profits of the top four oil companies alone were $100 billion in 2007 and Nobel Laureate Joseph Stiglitz in his book estimates the cost of the Middle East wars to be $3,000 billion or $3 trillion. In June, the International Energy Agency said that the world needs to spend $45 trillion to halve planet-warming carbon dioxide by 2050. We have a growing problem because it's hard to focus on Peak Oil and Global Warming when oil is

---

106 http://www.huffingtonpost.com/patrick-takahashi/trillions-and-quadrillion_b_156412.html

107 http://www.huffingtonpost.com/patrick-takahashi/billions-and-trillions_b_115491.html

108 http://www.huffingtonpost.com/patrick-takahashi/billions-and-trillions-re_b_134702.html

selling for $40/barrel and it is freezing outside. According to some reports, Europe and other parts of the world are experiencing record low temperatures. Remember. too, that my most recent post[109] feared that no vital decisions would be made on these issue because there is no present energy or environmental crisis.

Well, anyway, summer is still to come, who knows what next shock looms and we are beginning to enter new territory: quadrillion dollars. A Katrina victim is suing the U.S. Army Corps of Engineers for $3 quadrillion. NASA once reported that an ounce of anti-matter would cost $2.3 quadrillion to manufacture, if they could do it. All this is meaningless, of course, but five years ago, CNN.com reported that global warming solutions could cost up to $18 quadrillion, which was the highest end determination of expenditures required by 2100 of the United Nation's Intergovernmental Panel on Climate Change in 2003. But who takes them seriously? Not the G8 nations.

There is, though, at least one almost official sounding sum from the Bank of International Settlements: $1.144 quadrillion, which was the amount of outstanding derivatives worldwide at the end of last year. That would be equal to about $170,000 for each of the 6.7 billion people on Planet Earth. Mind you, this is funny money, for the current value of all the stocks and bonds around the globe is less than one tenth of a quadrillion dollars. Derivatives are unregulated and high-risk credit bets. Think Bernard Madoff. Warren Buffett, in a 2003 Fortune magazine issue, called derivatives the equivalent of financial weapons of mass destruction, and, surely enough, they actually exploded this fall.

But a quadrillion is 1,000,000,000,000,000--and only has 15 zeros. I calculated in **SIMPLE SOLUTIONS for Planet Earth** that the odds of any of us being born were 1 chance in 1 with 34 zeros. A googol has 1 followed by 100 zeros. There are many stories, but the most accurate one is that the founders of Google misspelled googol, and, anyway, google.com was available, and googol.com was not. For the record, the largest number with a name is the googleplex, which is one followed by...bah, this is getting too scientific. So we have a long way to go before our economy runs out of numbers.

Comments (13): The readership generally found this piece to be entertaining. I did, too.

---

109 http://www.huffingtonpost.com/patrick-takahashi/the-sun-will-continue-to_b_155160.html

This was a period when I worried that oil prices would continue to dip, thus wiping out any developmental interest in all the sustainables. I thought on 15January09 that the timing was now to immediately insert two important investment options (call them taxes, actually). Well, we missed the boat, as Saudi Arabia and the OPEC nations found a way to jerk the price back to the $75/barrel range.

Do It Now[110]

The signs are dismal. Yes, the economy, but worse regarding the dual hammer of Peak Oil and Global Warming. First, oil will apparently remain under $60/barrel for several years. Peaking at $147/barrel just this past July, Goldman Sachs expects oil to drop to $30/bbl this quarter. I must warn you that they did predict crude would rise to $149 by the end of last year, missing by about $100/bbl. But our U.S. Department of Energy predicted that oil prices would average $43.25/barrel this year (it was $99.55/bbl last year), and further, projected that the price will be $54.50/bbl next year. That agency has historically missed the mark with uncommon regularity. What if, though, GS and the USDOE are right this time? On top of all this, Secretary Designate Steven Chu, at his confirmation hearing, was picked up supporting coal and nuclear power. Yes, he had to balance his views for the Republicans, but the media emphasized that coal-nuclear point.

Second, millions probably won't perish from a terribly hot summer anytime soon. Well, that's good, but the problem compounding all this is a lot of misinformation being reported, including reports that the planet has cooled since George Bush has been President. The media likes to play with these stories, confusing the masses on something that still results in really cold winters and sea level rise that cannot be appreciated, especially if you happen to live in Kansas or D.C.

What are so bad about relief from climate change and LOW gasoline prices? Decision-making! The lack of crises will only further delay the need to take vital steps regarding energy and the environment. And time is of the essence.

Over the past few years, the United States has annually used about 100 quadrillion Btus of energy. Let's toss away that dimension (but analyze in terms of a hundred), so of that sum, 40 was oil, 23 coal and natural gas each, 8 nuclear and 7 (a bit high, but it was 6 in 2003 and looms to hit 7 soon) renewable energy. Biomass was 3.6, hydropower 2.5 and everything else was noise, although wind power is up to 0.32 and rising. Note that our energy consumption has not gone up much (98 in 2003), but, surprisingly, renewables have barely increased a grand total of one (1) in that period, partly because hydropower has declined. Yes, only a one percent improvement since 2003. Part of the reason for this lackluster advancement is the lack of enthusiasm for needed change.

---

110 http://www.huffingtonpost.com/patrick-takahashi/do-it-now_b_158133.html

But, unless we take decisive action now, we will have missed another opportunity. For the sake of Planet Earth and Humanity, we must immediately enact two meaningful measures, a national one and the other international.

1. First, Congress must add a $1/gallon gasoline investment surcharge now. We moaned at $4/gallon gasoline when Europeans were paying more than $10/gallon. The national price for premium today is less than $2/gallon. A whole dollar only kicks the price up to $3/gallon. This fund would accrue about $125 billion/year. Work out the rebates and subsidies, but spend most of the revenues on renewable energy research, demonstration and commercialization.

2. On a global scale, the G8 nations need to agree at their 2009 Italian Summit on a 1 cent/pound carbon dioxide remediation duty linked to crude oil at $30/barrel, to proportionately increase such that the levy would be 5 cents/pound carbon dioxide at $150/bbl crude. The sum collected could amount to about $200 billion/year just in the U.S. Funds should be used by each country to install smart grids and accelerate renewable energy investments. At 2 cents/pound, would almost double coal-fired electricity, making unnecessary any renewable energy tax credits, bringing into competitiveness wind power and, perhaps too, utility scale solar thermal power.

But, as explained in my ***HuffPo*** of January 5,[111] nothing will happen because there is no crisis. A potential fatal flaw of our society is the inability to make tough decisions absent any man-made or natural calamity. No doubt a tepid cap and trade attempt will be made and there will be sincere bluster about a workable replacement for the Kyoto Protocol. But those border on paralysis by analysis.

This $325 billion judgment palls in comparison to the Wall Street rescue package ($700 billion), the upcoming Obama stimulus program ($1 trillion) and the Middle East Wars ($3 trillion). Yes, this is comparing apples and oranges, for one, people hate taxes and can almost accept loans or fight terrorism, especially when you don't feel the daily bite. Nevertheless, this Peak Oil/Global Warming fund will insure for stable energy prices while providing development insurance to avoid an impending cataclysm[112] far worse than our current relative hangnail.

The American public believes, more than 90%, for example, in an ethereal concept such as a God. Is it possible for the Obama Administration to believe in the overwhelming evidence provided by scientists and specialists about the impending doom being forecast from the crush of Peak Oil and Global Warming, and, thereby, take immediate and appropriate action? Yes they can, so do it now!

Comments (2): The silence was deafening.

---

111 http://www.huffingtonpost.com/patrick-takahashi/the-sun-will-continue-to_b_155160.html

112 http://www.huffingtonpost.com/patrick-takahashi/the-venus-syndrome-revisi_b_140182.html

I chaired the Wind Energy Division of the American Solar Energy Society in the Mid-70's. In those days I did not have much confidence in this option. Boy was I wrong, as windpower today, with geothermal energy, is the only renewable energy technology reasonably close to competing with coal and nuclear. Thus, my posting of 26January09.

Where is Wind Energy Today?[113]

Wind and geothermal energy are the only competitive sustainable resources today, although a case can be made that wind power would not be able to produce electricity as profitably as coal or nuclear power facilities were it not for those tax breaks. Fortunately enough, the House Ways and Means Committee did not again hold the industry hostage until the eleventh hour when on January 22 it approved extension of the production tax credit for windpower[114] through 2012. Geothermal and solar incentives, if the legislation passes, would now be good through 2013.

In any case, if you were able to bring into the analysis the dangers of global warming and terrorism, then, arguably, these renewable options would be "competitive." But solar photovoltaics still have a cost problem and ethanol from biomass appears to be losing some support. Plug-in cars remain in the future with respect to meeting the test of clean energy, for most of the charging capabilities will need to be supplied by coal-powered generators for a long time to come. Remember that non-hydro renewables supply way less than one percent the electricity currently generated in our country.

This is not to say, however, that we should ignore these other solar options. The fact of the matter is that it will take time and nurturing (in the form of tax benefits and R&D) to get there, something we should have initiated after the 1979 second energy crisis. Energy from our winds, though, is at the threshold of being commercial.

I covered the story of wind energy in Chapter 2 of **SIMPLE SOLUTIONS *for Planet Earth*** (Everything you want to know about this subject can be found in the July 2008 publication of the U.S. Department of Energy entitled 20% Wind Energy by 2030.[115] I was particularly impressed with the group putting together this publication and the list of reviewers.) I first found myself in this field (details covered in the previously mentioned chapter) more than 35 years ago as chairman of the Wind Energy Division of the American Solar Energy Society. Many of the people from those days remained with the field and took part in this study.

In short, the report indicates that 20% of the national electrical supply could be met with wind energy by 2030. No technological breakthroughs are needed and 46 states would significantly contribute to the production. Half a million jobs would be created and annual property tax

113 http://www.huffingtonpost.com/patrick-takahashi/where-is-wind-energy-toda_b_160518.html

114 http://uk.reuters.com/article/idUKTRE50L7BC20090122

115 http://www1.eere.energy.gov/windandhydro/pdfs/41869.pdf

revenues should increase by $1.5 billion. A sum of $43 billion will need to be invested, but, and this is interesting, fuel expenditures would be reduced by $155 billion.

As 40% of total national carbon dioxide emissions come from electrical plants, the effect of wind power would effectively result in zero growth of this greenhouse gas from electricity generation by 2030. Further, water consumption from electrical facilities would be reduced by 17%.

There will need to be some understanding from environmentalists, for these wind energy conversion systems (WECS) can affect the life of a bird flying into the rotor, and resort hotels have been known to protest siting a wind farm nearby for fear of noise, visual pollution and image (the connection to the region being too windy). We faced these problems when we first introduced these systems a quarter century ago. One 3.2 MW device from Boeing, which had blades from tip to tip longer than a football field, built in the hills above Turtle Bay on Oahu, was particularly monumental. After a spell when only up to 250 KW WECS were being marketed, the latest devices are now re-entering this multi-megawatt range, for a 7 MW wind turbine was last year installed in Germany. This one takes 5 seconds to make a complete turn, so the bird count should not be a concern. Apparently, the problems we faced with materials and gearing have also been solved.

Hawaii, appropriately enough, has one-upped the nation, with a 40% by 2030 announced goal for renewable electricity. David Murdock himself has indicated an intent to spend $750 million to produce 400 megawatts of wind energy from the island of Lanai. There is a problem with who will be paying for the $1 billion or so undersea cable, but the resource is there and the need will come to a crushing reality within the decade, just about the time when everything should be in place.

Mind you, this comprehensive national effort for wind energy was initiated by the Bush Administration, so one can dream about 40%, too, or higher, from the Obama White House. Then, again, July of 2008 was when oil hit $147/barrel, and if current Department of Energy projections that crude will remain below $60/barrel for the next few years turn out to be true, then even the 20% by 2030 goal could well be a pipedream. Even T. Boone Pickens, with all his fanfare, essentially dropped out of the competition, for now.

The reality, though, is that the timing is perfect for President Obama and the Congress to actually, pass legislation for a $1/gallon gasoline tax and, perhaps, too, a one cent/pound carbon dioxide tax, for the consumer will hardly see the increase today. For example, the national price of gasoline hit $4/gallon in July and is now below $2/gallon, meaning that a $1 gallon tax only brings this cost up to $3/gallon, when double that amount is currently already being paid in Europe. Two cents/pound carbon dioxide tax would almost double coal fired electricity from 4 to 8 cents/kWh, making next generation solar thermal facilities competitive. My most recent **HuffPo** on "Do It Now" provides details.

Comments (14): There was good discussion, but most comments just nit-picked.

Towards the end of January I remained concerned that the still low cost of crude oil would undercut the long term development of renewable energy. I saw this in 1982 and again in 1998. Thus, on 29January09 came:

Cheaper Oil Will Improve the Economy, but...[116]

...deleteriously influence investments in next generation renewable energy systems. The U.S. uses about 20 million barrels/day of oil. In 2008, the average price of this oil was close to $100/barrel, or $730 billion/year.

However, the average price of oil is predicted to be about $43/barrel this year.[117] Thus, if true, we will be paying only $328 billion, saving $402 billion, which then, effectively, will enter the U.S. economy. How this multiplies into lower gasoline, airline tickets, food, fertilizer, and so on is beyond my analytical capability, but there must be a multiplier effect that increases the true windfall. Thus, in addition to the upcoming Obama stimulus package seemingly headed towards $1 trillion, just on lower oil prices alone, there will be another $402 billion sum provided to individuals, companies and governments. There should probably be a meaningful multiplier, maybe sufficient to bring this unexpected benevolence to a second trillion dollars, which will again be largely repeated in 2010 because the U.S. Department of Energy predicts that oil prices will then average $54.50/barrel.

Thus, the effective stimulus of more than $3 trillion (and up to $4 billion when you add on the Bush rescue package) over the next two years will probably boost the U.S. economy well beyond expectations. Similar advances will occur world-wide. The only ones to suffer will be those oil-exporting nations. It will be interesting to see what imaginative solutions OPEC will foist on the world to again embarrass the U.S. Department of Energy, which, granted, has been woeful in predicting the future price of petroleum.

Further, about two-thirds of the oil we use is imported. Thus, there is also the matter of $220 billion that we will not be paying to foreign oil suppliers in 2009. Somehow, there is then another nice multiplied sum you can add to the above each year until the anticipated next oil shock happens. Finally, any reduced Middle East war costs will further provide enhancement. Even now, we might be up to $5 trillion over the next couple of years to pump up the national economy.

On the above conditions, some optimism can be expressed about our future:

1. By the end of next year, the Dow Jones Industrials, which could, of course, slide a bit over the next few months, could well be 50% higher than today.

---

116 http://www.huffingtonpost.com/patrick-takahashi/cheaper-oil-will-improve_b_162019.html

117 http://www.bloomberg.com/apps/news?pid=20602099&sid=arHmrVb7s0Sc&refer=energy

2. Also in that time frame, don't be particularly surprised if crude oil crashes through $100/ barrel.

Yet, as the Sun will rise again tomorrow, you can expect skyrocketing crude prices, if not next year, then, surely, sometime within this coming decade. After all, our Energy Informational Agency has been regularly imperfect. In the meantime, as energetic as President Obama has been in his support for renewable energy -- surprising because with oil below $50/barrel, you would think that his foci would be limited to the economy and the Middle East -- with generally reinforcing comments from the Congress, I can only fear that the anticipated withdrawal of major industrial investments in sustainable resource projects will result in only admirable, but insufficient, action from our Congress.

To repeat, from my *HuffPo* on "Do It Now",[118] Congress needs to take advantage of the moment and immediately add a $1/gallon gasoline investment surcharge. We moaned at $4/ gallon gasoline when Europeans were paying more than $10/gallon. The national price for premium today is less than $2/gallon. A whole dollar only kicks the price up to $3/gallon. This fund would accrue about $125 billion/year. Work out the rebates and subsidies, but spend most of the revenues on renewable energy research, demonstration and commercialization.

Secondly, on a global scale, the G8 nations need to agree at their 2009 Italian Summit on a 1 cent/ pound carbon dioxide remediation duty linked to crude oil at $30/barrel, to proportionately increase such that the levy would be 5 cents/pound carbon dioxide at $150/bbl crude. The sum collected could amount to about $200 billion/year just in the U.S. Funds should be used by each country to install smart grids and accelerate renewable energy investments. At 2 cents/pound (when oil reaches $60/barrel), coal-fired electricity would almost double, making unnecessary any renewable energy tax credits, bringing into competitiveness wind power and utility scale solar thermal power.

What are the prospects of a $1/gallon gasoline investment surcharge and a one cent/pound carbon dioxide remediation duty (both, incidentally, also known as taxes) becoming real over the next year? One or more of you reading this posting could well be that galvanizing spirit to make this happen. Just read my Epilogue in *SIMPLE SOLUTIONS for Humanity* and be inspired.

Comments (3): There was one strong support from blog.cleantechies.com. Otherwise, nothing.

---

118 http://www.huffingtonpost.com/patrick-takahashi/do-it-now_b_158133.html

On 30January09 I followed with one of my more influential postings. I've long felt that the 4 cents/kWh cost of coal and nuclear electricity would be difficult to undercut. Then, I learned that new nuclear powerplants and coal with carbon capture were very expensive propositions. This changed the attitude of a lot of people, including myself.

Renewable Electricity is Our Only Viable Option[119]

First, a quick tutorial (details can be found in Chapter 1 of *Simple Solutions for Planet Earth*):

1. In reference to producing electricity, there is really dirty coal (remember acid rain and the Clean Air Act?), dirty coal (what is largely the practice today) and clean coal (nothing has really worked yet, but give them time). Further, along two tracks, there is the "clean" system that mines for coal but attempts to remove the carbon dioxide from the stack gas and store it underground, and the longer term option, which I worked on more than three decades ago at the Lawrence Livermore National Laboratory, called *in-situ* coal gasification (and oil shale retorting), where everything happens *in-situ* or in place. This technology was appearing to make a comeback when oil shot pass $100/barrel, but has been understandably quiet as of recent. You think we have a lot coal? Yes, we do, but we are also the Saudi Arabia of oil shale.

2. There are two types of nuclear power: fission (think Atomic Bomb), the currently utilized process, and fusion (Hydrogen Bomb and our Sun), the so-called cleaner and safer form, which in all pathways, seems to be a decade away from breakeven, as it has been since I worked on this concept in the 1970's. Thus, the engineering and economics are at least a generation away from commercialization, unless something like heavy ion fusion can suddenly gain credibility.

What will new nuclear and clean coal-fired electricity cost? If you go to traditional fossil and nuclear sources, you will see prices between 2 cents and 3 cents per kilowatt hour. So, as the average selling price of electricity today is 10 cents/kWh, why don't we just build more of these facilities? Well, for one, that's only from old facilities. Today, there are concerns about global warming and nuclear waste/terrorism. A whole new set of requirements needs to be met. Thus, it turns out that there will almost surely be a much bigger problem: **Economics**.

Consulting recent studies, projected electricity costs from new nuclear and coal plants seem to have jumped by a factor of at least three and as much as ten:

---

119 http://www.huffingtonpost.com/patrick-takahashi/renewable-electricity-is_b_162435.html

1. Joseph Romm earlier this month reported the cost of electricity from new nuclear facilities at from 25 cents to 30 cents / kWh, about triple the current price of electricity in the country, citing the study of Craig Severance.[120]

2. Romm also said last summer that the California Public Utilities Commission[121] placed the cost of power from new nuclear plants at 15.2 cents per kWh. They also put the cost of coal gasification with carbon capture and storage at 16.9 cents per kWh.

3. In mid-2007, a Keystone Center nuclear report,[122] funded in part by the industry, estimated capital costs between $3600 to $4000/kW, including interest. The report noted that the production cost would be 8.3 to 11.1 cents/kWh. In December 2007. Retail electricity prices then averaged 8.9 cents/kWh, so there would be no profit.

4. In October 2007, Florida Power and Light, a leading nuclear utility, presented its detailed cost estimate for new power plants[123] to the Florida Public Service Commission, concluding that two units totaling 2,200 megawatts would cost up to $8,000 per kilowatt, more than double that reported in the Keystone Report.

Why have these nuke costs escalated so much? Time (takes ten years from announcement to operation, and therefore, uncertainties about funding), fickle fuel fees for uranium, environmentalists and negative public sentiment, higher cost of materials and labor, waste storage nightmares, fear of terrorism, and more.

So from "too cheap to meter" to something anywhere from 9 cents/kWh to 30 cents/kWh, potential new nuclear power electricity rates now rest somewhere between solar photovoltaic and solar thermal electricity. Wind power is at half those costs, or even lower, and is sufficiently below the average cost of electricity to be on the cusp of being competitive with conventional coal. There is, though, that wheeling requirement that could double the cost of many windfarms.

Now, if global warming is real and coal-fired electricity with carbon capture/storage and new nuclear facilities are both in the conservative range of 15 cents/kWh, the solution becomes obvious: ***abandon building any new uranium/plutonium and coal power plants, and install as many wind farms and residential and utility-scale solar thermal systems as fast as possible.*** Also toss in geothermal energy into this mix. Assist in the promised coming of solar photovoltaics, and certainly, accelerate research into ocean thermal energy conversion, for this is the only baseload marine option of major promise. Offshore wind energy conversion systems are also beginning to show potential. I worry about wavepower, but there is hope for tidal and current power at a few choice sites.

---

120 http://www.grist.org/article/Exclusive-analysis-Part-1/

121 http://www.grist.org/article/life-after-coal/

122 http://www.nuclear.gov/pdfFiles/rpt_KeystoneReportNuclearPowerJointFactFinding_2007.pdf

123 http://energycentral.fileburst.com/EnergyBizOnline/2008-3-may-jun/Financial_Front_Prices.pdf

If global warming is a true concern, **economics alone** can justify this renewable electricity pathway. If you disagree, let me hear of your better solution. In any case, the more difficult problems are associated with developing sustainable fuels/systems for ground and air transportation. Several of my earlier *HuffPos* have addressed this challenge.

Comments (28): Well, this posting jarred a hornet's nest, especially many who were pro-nuke supporters. One of my responses was:

I don't want to sound like a raving anti-nuke maniac, because I have worked on fusion at the Lawrence Livermore National Laboratory, but here we have different opinions on fission, and I respect you on your position. My point of view is that we won't need to worry much about conventional nuclear power making any serious inroads because, if the utilities and public utility commissions are right, these facilities have too much risk and will be too expensive to build. As an aside, one of projects I help kill when I worked in the U.S. Senate nearly 30 years ago was the Clinch River Breeder Reactor. This happened not long after the Three Mile disaster, so it was an almost natural repercussion. I've wondered since then if that was a smart move, but, anyway, as I indicated, nuclear power plants will just be too expensive to build without heavy government support, and I don't think that the Obama Administration will give more than lip service to that possibility.

Here is a second response (to a PV comment), which sort of explains why I have been especially active during this period of relatively low oil prices:

Some day, I expect this to occur. The problem today is that solar photovoltaics remain a bit, if not a lot, too expensive for the normal household. Germany mandated that residential owners can sell their electricity back to their utility for more than 50 cent/kWh, so, rather suddenly, this country became #1 in the world. But Germany is not a particularly sunny nation, and this policy is under fire and has a sunset clause. While I strongly feel that we sort of blew it after the second energy crisis in 1979 when we did not put on a full court press to do as you suggest, the reality is that oil/gasoline prices dropped to an all time low (in current dollars) in 1998, and would do so again if crude drops below $30/barrel this year. Fickle energy prices have regularly destroyed good intentions.

One of my gripes in life is that we have lost the pioneering spirit. At one time other countries copied us. Today, we seem to be heading on a road to Japanese batteries. They have all the important patents. Let us head in a direction where we are dictating the technology. The direct methanol fuel cell, I think, is our best chance at building something better. The following was posted on 11February09.

America, Don't Copy...Build Something Better![124]

About a month ago I wrote an article[125] in **The Huffington Post**, wondering if there was a better option than the plug-in electric vehicle. Evidence indicated that the direct methanol fuel cell (DMFC) was worthy of exploration. To quote from that posting:

*Per unit volume, a fuel cell should be able to provide five times more energy than the lithium battery. Chapter 3 of **Simple Solutions for Planet Earth** can be referenced for details. In short, this device works like a battery to produce electricity, but uses hydrogen as the energy source instead of lithium, lead or cadmium. However, and this defies common sense, one gallon of methanol has more accessible hydrogen than one gallon of liquid hydrogen. So as hydrogen is very expensive to manufacture, store and deliver, with no existing infrastructure...*

...the logic argues for producing methanol from biomass to power a fuel cell. This simplest of alcohols is the only biofuel capable of directly and efficiently being utilized by a fuel cell without passing through an expensive reformer.

There were 32 comments, the most significant had to do with the glaring fact that there is no DMFC available for ground transport, and might well be impossible to build. Feedback from experts in the field, though, overwhelmingly indicated that, first, the DMFC is real, and could in a few years begin to replace batteries in portable applications such as computers and iPods. One obvious reason why there is no DMFC for cars is because the U.S. Department of Energy has prohibited R&D on this option. The Farm Lobby effectively convinced Congress and the White House to only focus on ethanol and biodiesel, purposely leaving out methanol. Details are provided in the above posting.

So therefore, where are we as a Nation on next generation cars? Not unlike our ethanol fiasco, the plug-in electric car has a different kind of gigantic problem: we don't produce any advanced batteries for this application. Panasonic and Sanyo[126] manufacture nearly all of the nickel metal hydride batteries used in our current hybrid autos. The irony to all this is that Stanford Ovshinsky of Ovonics in Detroit invented this technology, and even succeeded in suing Panasonic for stealing his idea. Ovshinsky worked out a partnership with General Motors, but their conventional wisdom must have prevailed, for nothing much happened.

---

124 http://www.huffingtonpost.com/patrick-takahashi/america-dont-copybuild-so_b_166134.html

125 http://www.huffingtonpost.com/patrick-takahashi/is-there-an-option-more-p_b_150824.html

126 http://featured.matternetwork.com/2008/12/short-supply-american-made-electric.cfm

The story of the lithium battery is also a national nightmare. If you trace the expertise, you will find yourself in Japan, Germany, South Korea, France and China. Where are the American companies with the world patents and leadership? They don't exist!

But, ah, we have coming the ballyhooed Chevy Volt. This hybrid will use a South Korean lithium battery and cost $40,000 when it becomes available in the fourth quarter of 2010. The Toyota Prius ranges in price somewhere in the twenties. The Volt is the car that will save Detroit?

We need to instead invent our own new power system. Toshiba and a few other Japanese companies do have a current advantage for portable uses of the direct methanol fuel cell (DMFC), but no one is doing anything about using this technology for cars, yet.

How ideal and opportune, then, for Detroit and the Obama Administration to partner on a new pathway for our future: initiate an Apollo-like project to develop the DMFC. The heartland of our country can also become involved, for the non-food portion of our crops and fields, cellulose, is the ideal feedstock for biomethanol.

The traditionalist might say, isn't this risky? It will take another decade or more just to build a competitive DMFC. Yes, they are right, it will take some time. But ten years from now, if we maintain our current course, we will be importing foreign batteries or paying royalties for our domestic brand. Is this smart?

The simple solution comes in two parts. Sure, continue the thrust to combine 14 American companies in a billion dollar federally funded venture to design an advanced battery.[127] This will almost surely be a lithium battery. Even though this has already been done, that's all right as a necessary hedge, for the plug-in concept has merit as a transition option. For one, the electricity can begin to come from wind energy.

The parallel focus should be to provide an equal sum to a consortium of American firms to accelerate the prospects for a direct methanol fuel cell. An important part of this effort should be to find a substitute for the platinum electrode, as, for example, carbon nanotubes. The potential is at hand to again become the world leader in vehicle production.

This second challenge is not currently being discussed in the White House or the Congress or Detroit. Why copy the world? Let us invent our own future.

Comments (8): My critic on the direct methanol fuel cell returned with vengeance. He could be right, but he is touting a different kind of fuel cell. I was disappointed that the public at large did not leap on the concept of Detroit and the heartland of American working together.

---

127 http://online.wsj.com/article/SB122957206516817419.html

Oil prices remained relatively low, and on 12February09 I again suggested that we take advantage of this period to enact a meaningful tax on gasoline and energy in general. Of course, this concept has zero chance today. So, I guess, we will await the jump and wonder why we did not do anything about this gift of time. Knowing those predictions below by the Department of Energy and Morgan Stanley, now you know why I don't have any confidence in them. They will be so far off that I wonder why they bother to keep embarrassing themselves. It turns out the Break Even Price of oil was the determining factor of future oil prices.

Will Oil Prices Remain Low for Two Years?[128]

Most probably yes, which will significantly affect the cost-effectiveness of most renewable energy projects, but perhaps not, as we shall see. First, what is low? Let me arbitrarily say low is $55/barrel (which is $1.30/gallon). Pardon me for this very tedious posting, but that's the way it has to be if you want to understand what is really happening.

There are 161 different crude oil prices. The New York Mercantile Exchange (NYMEX) version, linking West Texas intermediate grade crude with point of delivery at Cushing, Oklahoma, is usually the one mentioned in newspapers. The European counterpart is the Dated Brent Spot, and the difference between the two reached $9/barrel in mid-January. On 5Feb09, the NYMEX was at $40.85/bbl and the Brent at $46.46/bbl. This difference is not that important to this discussion, except to point out that all these prices vary quite a bit from each other.

Remember that oil went up to $147.27/barrel on 11July08 and dropped to $32.40/bbl on 19December08, a four year low. This is not close to the $11.91/bbl of 1998, or about $16/bbl in current dollars. In 1972, before the first energy crisis, oil in current dollars sold for $23/barrel (actual price then was $2.85/barrel), or higher than in 1998. This is significant. Let me repeat: the cheapest oil has ever been in current dollars was only eleven years ago.

On January 1, 2009, the Organization of Petroleum Exporting Countries (OPEC) reduced output by 2.46 million barrels/day. The world used 86 million bbl/day in 2008, will drop closer to 85 million bbl/day in 2009 and will probably return to 86 million bbl/day in 2010. So the reduction only amounted to 3%. In comparison, the U.S. is producing at a rate of 5 million bbl/day and consuming at 20 million bbl/day.

What then are the oil prices to come? The U.S. Department of Energy[129] has projected an average of $43.25/bbl in 2009 and $54.50/bbl in 2010. Morgan Stanley[130] predicts a drop to $25/bbl in the second quarter and an average of $35/bbl in 2009, more than $8 lower than the USDOE, with an increase to $55/bbl in 2010, about the same as the USDOE.

---

128 http://www.huffingtonpost.com/patrick-takahashi/will-oil-prices-remain-lo_b_164901.html

129 http://www.bloomberg.com/apps/news?pid=20601207&sid=arHmrVb7s0Sc&refer=energy

130 http://www.energyandcapital.com/articles/oil-price-forecast+2009/815

Tossing these speculations into a pot and stirring, it looks like oil will drop below $30/barrel by mid-year, and average about $40/bbl this year. Next year looks to be $55/bbl. Again, though, oil at $147/bbl caught the USDOE by surprise (possibly because their system does not allow for sudden re-predictions), while Goldman Sachs was looking at $200/bbl in the near future and was off by a factor of three when their recalculated ballpark guess of $100/bbl for December 2008 sunk to $32.40/bbl instead. My rule of thumb is that these prognosticators could be right, but they are usually wrong.

What is the reality of average oil prices remaining at or below $55/barrel for a couple more years? Well, it should depend on supply, demand and the state of the economy, right? But if you were an oil producer, here are the realities:

1. The break-even price (BEP, anything they sell for less than this price, they lose money) of oil, as indicated by the International Monetary Fund (IMF),[131] for Iraq is $94/bbl (or $111),[132] Iran $90/bbl[133] and Venezuela $58/bbl. Are you already uneasy?

2. The National Bank of Kuwait[134] reports that the BEP for Saudi Arabia is $30, but IMF says $54/bbl and Wikipedia[135] lists $49/bbl.[136] I would believe something closer to $50/bbl than $30/bbl. So, do you think Saudi Arabia could survive if they have a minus balance on oil for the next two years? No wonder that they regularly suggest $75/bbl as being fair to the world.

3. Canadian oil sands are reported by the National Bank of Kuwait to have a BEP of $33/bbl, but the Calgary Herald[137] indicates a figure over $54/bbl. Plus, there are those fierce environmental and global warming concerns dogging this resource.

So, on the one hand, you have the Department of Energy and Morgan Stanley officially pronouncing that oil will average something in the range of $40/bbl this year and $55/barrel next year. On the other, you have Iraq, Iran and Venezuela all with a break-even prices greater than $55/barrel. These three countries, and, probably Saudi Arabia, too, will be losing money through next year. Will that be tolerable? If all the reported information is accurate (even though they never really are), then I can only fear the worst. My gut sense is that political solutions will become necessary imperatives. You can let your imagination run wild on what those may be.

---

131 http://www.gulf-times.com/site/topics/article.asp?cu_no=2&item_no=242775&version=1&template_id=48&parent_id=28

132 http://www.freemuslims.org/news.php?id=4127

133 http://threatswatch.org/rapidrecon/2008/10/irans-fear-of-low-oil-prices/

134 http://seekingalpha.com/article/58322-oil-price-predictions-and-break-even-prices

135 http://en.wikipedia.org/wiki/Magic_number_(oil)

136 http://www.freemuslims.org/news.php?id=4127

137 http://www.calgaryherald.com/business/Crude+price+royalties+restore+hope+oilfields/3096940/story.html

However, you can also think positively and treat this period to be a Gift to Planet Earth and Humanity,[138] for time is being provided to develop the technologies and systems to combat Peak Oil and Global Warming. Secondly, let's face it, cheaper oil will improve the economy, for, combined with all the stimulus packages, $5 trillion dollars will thus be pumped into the economy. There is a but,[139] though, to this second gift. Will we be able to take advantage of this reprieve? We need to Do It Now.[140] What has to be done NOW? Click on that posting.

Comments (12): There was general agreement about my premises. Still, I'm afraid, no solutions to do something about all this.

---

138 http://www.huffingtonpost.com/patrick-takahashi/a-gift-to-planet-earth-an_b_148902.html

139 http://www.huffingtonpost.com/patrick-takahashi/cheaper-oil-will-improve_b_162019.html

140 http://www.huffingtonpost.com/patrick-takahashi/do-it-now_b_158133.html

The legal counsel for the Blue Revolution, Leighton Chong, succeeded in copyrighting BLUE REVOLUTION for me, and indicated that I needed to do something about it. This 16February09 posting will hopefully satisfy the government that I am serious about this all.

Blue Revolution[141]

A few months ago, my Huffington Post article on "The Dawn of the Blue Revolution"[142] reported on the potential the ocean provided for humanity. The world has been wrestling with the economic collapse and worried about the dual hammer of Peak Oil and Global Warming. Almost never is the ocean recognized as part of the solution. The surface of our globe has two and a half times more water than land. Surely, then, greater consideration should be given to this largely ignored and mostly protected portion of our planet.

To quote from the above posting:

In my 2003 Bruun Memorial Lecture to UNESCO in Paris, I proposed that the United Nations take a leading role in galvanizing Project Blue Revolution. There are important Law of the Sea and international political issues to be considered. There are today only 192 countries forming the UN. Someday, perhaps, a thousand OTEC-powered Blue Revolution platforms, each a nation in itself, could well be plying our oceans, providing clean and sustainable resources for Humanity in harmony with the ocean environment.

Well, the thought of a thousand new nations is disarming at best, but noteworthy because of the potential immensity of the promise.

Not too long ago we had the romance of space with Star Wars and the Apollo Project. But each NASA shuttle shot costs about a billion dollars, the same amount as the U.S. Department of Energy has annually spent over the past decade to develop renewable energy. We can keep contemplating the sky, but the reality is on Earth, and, more specifically, our seas.

Where else can we turn to for next generation clean energy technologies, green materials, exciting new habitats and more seafood? The ocean, of course, and three recent developments have particularly caught my attention.

First, I have never met him, but initiated some virtual dialogue with Patri Friedman, self-professed poker authority, who left **Google** to help form the Seasteading Institute (as opposed to homesteading on land).[143] Funding was received from billionaire Peter Andreas Thiel, chess-master and co-founder of PayPal. My only connections with them are that we all went

---

141 http://www.huffingtonpost.com/patrick-takahashi/blue-revolution_b_166977.html

142 http://www.huffingtonpost.com/patrick-takahashi/the-dawn-of-the-blue-revo_b_145889.html

143 http://seasteading.org/mission/intro

to Stanford University and have the belief that our oceans provide a hope for our future. They are dissatisfied with the current civilization and want to build a new and better one on the high seas. Their goal is to form new nations as seatopias.

Second, the Blue Revolution[144] was trademarked (77/452663) to spur the concept, which is similar to seasteading, except the latter is a whole new society, while the former is a focus on the technical aspects of optimizing a system powered by ocean thermal energy conversion[145] to support a floating city or industrial complex, with next generation fisheries,[146] marine biomass plantations for biofuels, hydrogen, green chemicals and materials and freshwater, while remediating global warming[147] and preventing the formation of hurricanes. Corporate interest has been expressed to carry on the effort.

The third blip of activity is the Japanese Ocean Sunrise Project.[148] The chairman of the group, Toshitsugu Sakou, last week passed on to me their final report on "Seaweed Bioethanol Production in Japan." Unfortunately, both volumes were in Japanese. However, the journal paper he gave me reported that their aim is to produce seaweed (a brown macroalgae) bioethanol by farming in the Exclusive Economic Zone of their country. They don't quite say that the ocean can produce all the fuel necessary for the world, but they did mention that the theoretical limit of terrestrial cellulose was such that only 18% of transport needs could be met. Clearly was the implication that the ocean can much better meet the full challenge.

As land-based societies attempt to cope with economic, social, political, environmental and assorted other problems, the notion of a totally new initiative to do it better this next time seems enticing at sea. Someday we will explore the stars, but for the next few centuries the ocean is our only new frontier for progress.

Comments (2): There was only one supporting response, plus mine. I would surmise that the Blue Revolution has not caught on because very few, indeed, are interested in this glorious pathway for humanity in harmony with the marine environment.

---

144 http://planetearthandhumanity.blogspot.com/2008/08/blue-revolution-in-hawaii-part-8.html

145 http://www.huffingtonpost.com/patrick-takahashi/the-coming-of-otec_b_145634.html

146 http://www.huffingtonpost.com/patrick-takahashi/the-ultimate-ocean-ranch_b_146192.html

147 http://www.huffingtonpost.com/patrick-takahashi/the-venus-syndrome-part-o_b_106120.html

148 http://ieeexplore.ieee.org/xpls/abs_all.jsp?arnumber=4449162&tag=1

On 20February09 I went back to exhuming the concept of every country decreasing their defense budget by 10% annually. I added a few more non-compelling points. Is there any hope for peace forever? No, not now, not ever...maybe.

The 10% Solution[149]

President Ronald Reagan jacked up the defense budget, and it worked, for the strategy bankrupted the Soviet Union, we won the Cold War, and we now have no major enemy. There is every good reason, thus, to drastically change our spending priorities. The threat of nuclear holocaust has shifted to the economy, peak oil and global warming.

Wait a minute, that was twenty years ago, and nothing has changed. Why don't we now smartly reallocate most of our defense spending to green applications? Boeing can make wind energy conversion devices, Lockheed Martin and General Dynamics manufacture OTEC plantships, etc. My very first *Huffington Post* article[150] suggested this plan to Barack Obama last year when he was still trying to gain the democratic nomination.

Here is one way of looking at this. The next nuclear carrier will cost $8 billion. That is about the total amount the U.S. Department of Energy spent on renewable energy R&D over the past decade. After the carrier is delivered, the lifetime operational cost could well be three to four times the capital cost, so we are now in the range of $30 billion to $40 billion. Even though aircraft carriers are strategically obsolete, for the next couple of decades, it will serve the purpose of intimidating minor enemies, so keeping the peace is worth something, I guess. Plus, very little carbon dioxide is generated.

But, say these tens of billions are utilized to stimulate the building of wind farms, solar power facilities, developing next generation transport options and the like. Clean energy will be produced, plus more and peaceful jobs will be created to better benefit humanity. Revenues will be generated, the economy will be sparked and our atmosphere cleansed. This should be a no brainer.

Unfortunately, this did not happen in 1989 and won't in 2009. Why won't our decision-makers take such an obviously sensible step? American companies can dominate the competition for defense funds. They don't need to compete against those foreigners. They have effective lobbyists. Anyone in the know has no confidence that Congress can act for the common good on these matters. This is one of the glaring flaws in our political system.

The second reason is that the public likes to feel secure, and, as long as no American is being killed, it's okay, if not desirable, to have the best (and most expensive) weapons to protect our

---

149 http://www.huffingtonpost.com/patrick-takahashi/the-10-solution_b_168090.html

150 http://www.huffingtonpost.com/patrick-takahashi/well-barack-we-have-a-pro_b_104201.html

country, especially if the factories or bases are in their home state. The military industrial complex has planned wisely, for virtually every state does, indeed, benefit from this jobs machine.

That thinking should be obsolete today, but is not, for it takes time for a society to realize that we have no enemy anymore, and won't for a long time to come, if ever. There is that rag-tag bunch of terrorists, but China and Russia have their own problems, Iran is now losing money producing oil and North Korea is a joke. Further, as badly as the U.S. stock market did last year, we were still about the best in the world, dropping only 35% in value. Chinese and Russian stocks fell 70%.

Make no mistake about it, though, this recession is leaning in the direction of depression, and those trillion dollar rescue packages dwarf past government expenditures. Yes, World War II did cost something on the order of $2.5 trillion, but how many realize that the total of the Manhattan Project, Marshall Plan and Apollo Project, in 2009 dollars, is only about a quarter trillion dollars? The Bush and Obama rescue packages alone amount to five times more than what it took to build the atomic bomb to end the war, save Europe and send Men to the Moon.

That was bad enough, but the issue is, actually, larger than the country and beyond the economy. When you add the potential cost of combating Peak Oil and Global Warming, the challenge facing society becomes downright depressing, for the International Energy Agency last year reported that $45 trillion was the amount needed only to neutralize climate change. That's about forty times the Bush/Obama stimulus.

So, clearly, the challenges are monumental and solutions must be international. Forget the United Nations, for the bureaucracy is stultifying. If only one monumental agreement can be made at the next G8 summit, which will occur in July of this year in Italy, the world can begin to work together on a common solution for Planet Earth and Humanity.

What could they possibly agree on? Well, let me be brutally naive. If each country can reduce defense spending by 10% each year, with the resultant savings being applied to renewable energy and climate change, that would be a nice start. It will take time for each country to arrive at a legislated resolution, so the protocol for agreement can be signed in the 2010 Canadian Summit. By 2012, when President Barack Obama hosts the G8 gathering, the world economy should be well on the way to recovery, with oil prices tolerable, atmosphere perhaps already showing signs of healing and a globe at peace. Hey, what's wrong with a little loud dreaming? Maybe someone will take some of this seriously.

Comments (15): There were a lot responses, most very supportive. I quote myself:

We seem to generally agree in principle. Yes, much of this was tongue in cheek, but, as I said, who knows, maybe someone out there is reading this stuff and is in a position to take significant action. You need to venture to gain.

On 23February09 I posted, probably, my favorite HuffPo. The effort linked various topics covered by my two SIMPLE SOLUTION books, but was, well...funny. Several colleagues said they laughed out loud. No one particularly was influenced by what I was trying to say, but, all in all, a very satisfying experience.

Evolution, Global Warming, Doomsday and the Afterlife[151]

About that title, Carnac the Magnificent would pause because he would at least be mildly confused as to his possible response. In time, he utters, "Not the life progression for Americans" and opens the envelope. There is only a trickle of laughter because the audience does not quite understand the connection. Johnny Carson grimaces and extends a particularly invective Middle Eastern curse.

Yes, Carnac would have been right, for the above title is not a sequential prediction of your past, present and future, if you are the typical American. But shockingly enough, not for the reasons you might think. The punch line comes at the end when you learn that we (not me, but the majority) believe in only one, well, maybe two, of the following: evolution, global warming, doomsday and the afterlife.

I recently wrote a book partially treating evolution and the afterlife,[152] and an earlier one analyzing global warming and possible doomsdays,[153] and have lately been busy reading up, responding to and organizing virtual discussion forums on the economic collapse, peak oil, global warming and doomsday. The participants tend to be the more scholarly facet of our society and hardly represent the masses. Yet, many of them have devoted their professional lives to these subjects and are the best and brightest, so their views are important and can't be totally discarded.

I would say something like 10% to 20% of this eclectic group actually believe the combined crush of the economic collapse, peak oil and global warming will be so severe that society as we know it will NEVER recover, our lifestyles will be seriously compromised and survival could become a life-or-death issue. Some have purchased land and are initiating self-reliant communities. There are blogs to prepare individuals living in these uncertain times.[154] A few of my friends are actually looking forward to this new kind of adventure. Most in this forum, though, are like me, in that we rather enjoy our current mode of life, but are beginning to get mildly concerned.

---

151 http://www.huffingtonpost.com/patrick-takahashi/evolution-global-warming_b_168827.html

152 http://www.authorhouse.com/BookStore/ItemDetail.aspx?bookid=50982

153 http://www.authorhouse.com/BookStore/ItemDetail.aspx?bookid=46634

154 http://www.survivalblog.com/

I tried searching for a poll on what real Americans think about this coming doom, but couldn't find one with a scenario reasonably close to the above. Sure, there are the Biblical ones, and, well, we are a religious country, plus the Large Hadron Collider still has that miniscule potential and the SciFi Channel in 2006 had a countdown to our mass extinction.[155] However, as none of the respected economists or politicians I see on television seems particularly concerned about this worst-case option, I take satisfaction in maintaining a similar insouciance.

Anyway, how can anyone get so traumatized by this latest series of existing and potential catastrophes, for in the 70's we muddled through the population bomb, limits to growth, potential nuclear winter, the Vietnam War, acid rain and two energy crises...and somehow recovered. In fact, nearly two decades after the Second Energy Crisis, crude oil in 1998 fell to the lowest on historical record ($15.52/barrel in 2008 dollars, even lower than the $18.29 of 1972), and there was nothing government, academics or politics did to orchestrate this drop. Further, another decade later the United States is now supremely unchallenged and oil is heading back to almost historic lows. Nuclear holocaust? Iran and North Korea will not precipitate a World War 3. Aside for this inconvenient economic collapse, things seem generally okay today.

All this led me to think, though, whether this small minority planning for the end (of life as we like it), might, in fact, be right? Let's look at religion, for example. The surveys vary a bit, but for the longest time, something on the order of 90% of Americans have said they believe in God[156] and some form of Afterlife. Of that remaining 10% who don't, I would not be surprised if many of them are amused and disappointed at the same time that their friends and family can be so deluded. Is that what this doomsday group thinks of the population at large in terms of the coming downfall?

Changing the subject a wee bit, we are celebrating the 200th anniversary of Charles Darwin. By now, everyone must believe in evolution, right? Wrong! Two-thirds of Americans actually know that God created us within the past 10,000 years,[157] so does this mean that only 33% of us accept evolution? Well, it's a bit more complicated then that, but, yet, alarming. Live Science reports on a poll of 34 countries, placing the U.S. second from the bottom[158] (Turkey was lower at 26% on belief in evolution, while the European countries and Japan were just the opposite, with 60-90% in the evolution camp). I go into vivid detail on this subject in Chapter 5 of *SIMPLE SOLUTIONS for Humanity.*

Okay, so much for that controversial subject, but, then, by now most of us must be convinced about global climate change. Actually, yes, 82% of Americans do believe,[159] with, interestingly

---

155 http://www.nhne.org/news/NewsArticlesArchive/tabid/400/articleType/ArticleView/articleId/1108/SCI-FI-Channel-Countdown-To-Doomsday.aspx

156 http://www.christianpost.com/article/20070402/poll-9-of-10-americans-believe-in-god-nearly-half-rejects-evolution/index.html

157 http://www.usatoday.com/news/washington/2007-06-07-evolution-debate_N.htm

158 http://www.livescience.com/health/060810_evo_rank.html

159 http://www.foxnews.com/story/0,2933,250571,00.html

enough, 91% being Democrats and 72% Republicans. That was a 2007 survey. A 2009 poll broke this down to 44% saying long-term planetary trends are causing this change, with 41% blaming it on human activity.[160] Only 21% of Republicans think that we are at fault. Also, 54% say the media exaggerate the dangers. In other words, most Americans don't think their use of fossil fuels is causing this Greenhouse Effect. They blame nature.

To summarize, the majority of Americans believe in both creationism and an afterlife, the potential of some sort of religious doom, and think they are not causing global warming. So the title of this article should have been: "Creationism, Doomsday and the Afterlife," to more closely reflect life in the USA. You now should have a better understanding about why we are in deep...(feel free to add your own odious term). So what has this got to do with the economy? Go back to the beginning and try again, or revert to my earlier **HuffPost** introduction to this subject.[161]

Did we become the greatest country ever because of our beliefs? Certainly not entirely, which gives me hope that the best is yet to come.

Comments (10): Shockingly, to me, just about everyone agreed with me. To quote myself, in summary:

If anyone bothered to reread this article, then, still not sure how Evolution, Global Warming, Doomsday and the Afterlife affected the economy, went on to my earlier posting on this subject, and yet could not figure out the connection, welcome to the real world.

I taught a course called "Technology and Society" for a number of years in the seventies. We covered all the subjects mentioned above. The final essay question I generally gave, warning the student not to spend more than 5 minutes on it, was, relate the following three terms to the purpose of the course. I tried to pick three words that could not possibly have any connection. The responses were amazing. They almost all came up with sensible responses, with the poorer students being the most inspired.

In a way, the resolution of our current crises will call for creativity of this sort. The fact that that our beliefs are all over the spectrum can be a plus as we attempt to arrive at a solution. Then again, I wonder if a country like Singapore, which has a regimented mentality focused on realistic answers, might have a better chance at solving this puzzle. Perhaps this benevolent dictatorship model is something we might want to consider. At the terminal extreme, maybe Jesus will come.

---

160 http://climateprogress.org/2009/02/03/rasmussen-reports-global-warming-denial-gop-polling/

161 http://www.huffingtonpost.com/patrick-takahashi/global-warming-and-the-af_b_117385.html

I've long been wondering why America does not sign those international treaties. My three year assignment in the U.S. Senate gave me a glimpse of the politics involved. But we are notorious for not ratifying many humanitarian agreements virtually the entire world has signed. For the record, a recent HuffPo reported that the Obama Administration will not sign the land mine treaty. So my posting of 16March09 investigates why, and could not rationalize the reasons.

The USA: A Good World Citizen?[162]

In a word: NO! I was amused when I recently read that French President Nicolas Sarkozy announced that France will return to full membership of the North Atlantic Trade Organization (NATO).[163] What, they have not been a full member? It turns out that in 1966, President Charles de Gaulle withdrew his country from the alliance's command structure. So I wondered, what about the good ole U.S. of A? Are there any similar surprises?

In my **SIMPLE SOLUTIONS for Humanity**, I argued that the United States is the best country ever. I still believe so, but I have to wonder why we have been such a BRat (Bully Rat) in a variety of universal world agreements. I can tell you at the outset: *Conservatives* have generally been the reason why. Although I have sometimes tended to side with them, let me be specific.

First, the Kyoto Protocol. One hundred eighty three countries have ratified it, yet the U.S. remains the *only* major nation in opposition. Enough said.

Of course, many times we do not sign on for "good reason." Take global warming, for example. It makes little sense to do so when China and India are pardoned. However, International Women's Day was March 8, and I found it astounding that the Global Women's Rights Treaty,[164] completed 30 years ago, was endorsed by 170 countries, but not the USA. Why? There is some fear of promoting prostitution and abortion. Would you guess that, as in global warming, hardcore conservatives might be blamed? We are in abysmal company, for Sudan, Somalia, Qatar, and Iran have also refused to sign.

Senator Barbara Boxer (California-D), though, has taken on the challenge of gaining ratification. She has also added the United Nations measure to expand the rights of children.[165] It is reported we have dragged our feet on this issue for fear of condoning youth soldiers, children smoking marijuana, and the like. In case you missed the point, there is a sector of our society that believes parents should have the right to rear their own children, which, granted, makes some sense. This has been floating around for two decades, and was actually

---

162 http://www.huffingtonpost.com/patrick-takahashi/the-usa-a-good-world-citi_b_174904.html

163 http://news.bbc.co.uk/2/hi/europe/7937666.stm

164 http://www.sfgate.com/cgi-bin/article.cgi?f=/c/a/2009/03/07/MNJ916B87L.DTL&type=politics

165 http://www.foxnews.com/politics/2009/02/25/boxer-seeks-ratify-treaty-erode-rights/

signed by President Bill Clinton, but not approved by the U.S. Senate. New UN Ambassador, Susan Rice, recently said that this is a noble cause backed by the Obama Administration.

In case you were wondering, the President signs treaties, which then require Senate approval. The House does not get involved. Oh, by the way, Clinton also signed the Kyoto Protocol, but was not able to gain Senate support.

The Law of the Sea Treaty (LOST)[166] is, in additional to being an appropriate acronym, one that I have actually been a part of since my stint in the U.S. Senate three decades ago. While these LOST summits have been close to being the most boring of any possible human assemblage (although, maybe those UNESCO Intergovernmental Oceanographic Commission biennial gatherings exceed that mind-numbing quotient), you would think that something passed in 1982 to cooperate on ocean matters would be a slam dunk certainty. A combination of companies not wanting to share information and something to do with sovereign rights over the open ocean has served to scuttle our participation. Interestingly enough, President George Bush, the 43rd president, actually backed this longstanding treaty. Secretary of State Hillary Clinton, UN Ambassador Rice and Vice President Biden are in support, so surely it will be ratified this time, right? Well, stay tuned, for the military industrial complex suite of war partners remain adamant against this giveaway.

Okay, who can possibly be for land mines? Well, the USA. We have refused to join 150 nations that have ratified this treaty. This one is only a decade old, but the Pentagon is against it because signing would compromise South Korea. Huh? There is also a Cluster Bomb Ban[167] that the U.S. and Russia stand opposed to. Odds are that the Obama Administration will move on these, for they do have humanitarian and practical benefits.

The International Criminal Court[168] was formed in 2000 to investigate and prosecute genocide, crimes against humanity and war crimes. President Clinton previously signed, but did not even bother to submit it to the Senate for ratification. George Bush the Younger, then "unsigned" it in 2002 because it could bring politically-motivated prosecutions against U.S. citizens. It is possible that the U.S. Constitution would need to be amended to ratify, and there are more important measures to pass at this time, I guess.

There are a few more, but let me stop with the Comprehensive Nuclear Test Ban Treaty.[169] Certainly, we must be the good guys here. Nope. One hundred eighty one nations have signed on, but the US, China, India, Pakistan, Israel, Iran and North Korea have not approved the treaty. Russia is one of the 148 that signed and ratified.

---

166 http://www.letfreedomringusa.com/news/read/334

167 http://www.huffingtonpost.com/2008/12/03/over-100-nations-back-clu_n_148032.html

168 http://www.globalpolicy.org/empire/us-un-and-international-law-8-24/us-opposition-to-the-icc-8-29.html

169 http://www.espionageinfo.com/Co-Cop/Comprehensive-Test-Ban-Treaty-CTBT.html

Let me finally end this discussion with the UN Security Council and veto power.[170] Without going into the details, the Soviet Union vetoed more than 100 resolutions. Since the fall of the Berlin Wall in 1989 there were 23 vetoes, the U.S. with 15 and Russia with 6.

This is getting monotonous, so let me conclude with the simple statement that now you probably should have a better inkling as to why we have not been really popular around the world. Mind you, popularity is tertiary to being right and humane, but consider the above and let your wishes be known to our current leaders.

Comments (0): Yes, no response. I thought I would get conservatives out of the woodwork.

---

170 http://www.globalpolicy.org/security-council/tables-and-charts-on-the-security-council-0-82/subjects-of-un-security-council-vetoes.html

When I posted this article on 2April09, about 300 extrasolar (planets around other stars) had been identified. On 17June10, the number had increased to 461. People are fascinated about space, and ET movies in particular are popular. I once worked for NASA, but now feel that much of the current expenses are unwarranted. As we will not be involved with actual space travel for some time to come, my recommendation has been to continue a cost-effective scientific program. The Search for Extraterrestrial Intelligence, I think, is one such endeavor worthy of support.

Extraterrestrial Intelligence?[171]

America has a fascination for extraterrestrials (ET). *ET*, the 1982 movie, was the most financially successful film of all time released to that point. Thirty-two years ago -- can you believe it was that long -- Steven Speilberg wrote and *directed Close Encounters of the Third Kind. Monsters vs. Aliens*, **Knowing** *and The Race to Witch Mountain* were #1 at the box office the past three weeks. These flicks have entertainment value, but what is the reality of extraterrestrial intelligence (ET) and flying saucers?

First how do Americans feel? Remember, 90% of us believe there is a God and the majority does not think it is responsible for global warming.[172] There is a plethora of polls, but, in general, 60% feel there is ET life somewhere in the universe.[173]

Are there UFOs? Absolutely, for any flying object that cannot be identified is by definition a UFO. However, what about flying saucers piloted by some extraterrestrial? Anywhere from one third to one half of Americans think they are real. In comparison, one half to three fourths believe in angels, yes, the kind with feathered wings.[174] Religion is dominant in the USA.

Most scientists don't want to even be associated with UFOs because the odds are far too low to warrant their attention, not to mention the potential of sullying their credibility. Consider this: light, which in one second can travel around our globe 7.5 times, takes 100,000 years just to get from one end of our little ole Milky Way galaxy to the other. Us, Homo sapiens,[175] only appeared 100,000 years ago. Then, if you want to get to Andromeda, our closest galaxy (the latest guess is there are 500 billion galaxies up there), a spaceship traveling at the speed of light would take 2.2 million years. Plus, what form of energy will enable a craft to travel

---

171 http://www.huffingtonpost.com/patrick-takahashi/extraterrestrial-intellig_b_180768.html

172 http://www.huffingtonpost.com/patrick-takahashi/evolution-global-warming_b_168827.html

173 http://www.space.com/news/050531_alienlife_survey.html

174 http://www.foxnews.com/story/0,2933,99945,00.html

175 http://www.cartage.org.lb/en/themes/Sciences/Lifescience/PhysicalAnthropology/HumanGeneticEvolution/EarlyModern/EarlyModern.htm

such distances? Sure, you read about wormholes and stuff like that, but, for now, it is best to be rational and consider becoming a nonbeliever of flying saucers.

In Chapter 4 of **Simple Solutions for Humanity**, I relate my experiences in this field, starting with Project Cyclops, and also Orion, a short stint I had at NASA's Ames Research Center. The question then was, are we the only planet in the universe? I interacted then with Barney Oliver, Jack Billingham and Carl Sagan, and actually proposed a project to detect earth-sized planets. The concept rested on the principle that for life to occur, there needs to be an atmosphere, and starlight (sunlight) causes population inversion (a condition which induces lasing), meaning that spikes of monochromatic light can be detected, both proving that a planet exists and providing the gas composition. I took cues from Charles Townes, who had just moved from MIT to Cal-Berkeley and wrote on this subject in **Science**. NASA tossed my proposal aside and remarked that the Hubble Telescope would soon fly and will then accomplish this task. Well, earth-sized extrasolar planets are beyond the capability of Hubble.

So, a little more than two years ago, the European Union launched CoRoT[176] to find extrasolar planets similar in size to ours. The principle had to do with these planets transiting (moving across the star) and measuring any diminution of light. On March 6, NASA placed into space Kepler,[177] at a cost of nearly $600 million, to do almost exactly the same thing.

The key question I ask is, why do we need both? Secondly, as the atmospheric composition will not be determined through this copycat photometric technique, so the potential of life as we know it cannot be determined, for so much money, couldn't we have somehow adjusted the mission to provide more useful answers?

Okay, anyway, now we know that there are more than 300 planets out there somewhere. Currently, most "seen" (we have not actually seen them, we have mostly measured wobbles in stars, surmising that the movement must have been caused by a planet or more) have been Jupiter-sized or larger, but just the fact that there are other solar systems answers the original question: yes, there are a lot of planets around other stars.

Now that we have proven that there are probably billions of planets out there, with odds that some of them could, perhaps, have intelligent life billions of years more advanced than ours (the Universe is slightly less than 14 billion years old and our solar system has been around for less than 4.5 billion years), let us get on to detect possible signals, as Jodie Foster did in **Contact**. Yes, I know that was a movie, but this is a life posting, not a scientific publication.

Oh, by the way, did you know that historically, NASA was prohibited from doing SETI research? Yes, there is a privately funded SETI Institute (the Paul Allen Telescope Array is one of their projects), but Congress did not allow the federal government to directly spend money on the **Search for Extraterrestrial Intelligence**. That is, until 2003, when congressional

---

176 http://www.esa.int/esaSC/SEM7G6XPXPF_index_0.html

177 http://kepler.nasa.gov/

attitudes somewhat shifted, and NASA actually began to provide a few bucks to this activity, but only a very few.

In this time of economic turmoil, can funds be justified for SETI? If we can spend $600 million for Kepler when Europe already was doing that, will expend $8 billion for our next nuclear aircraft carrier (which is by any current war standards already obsolete), and provided $150 billion of bailout money to AIG, sure, a justifiable amount would be worth the investment, for, perhaps, streaming in from space could be the answer to world peace, cure for cancer, solution for global warming and resolution to our global financial crisis.

Our civilization will survive recession, Peak Oil and Global Warming, as we did the Cold War. Can our next few billion years, though, be more progressive and successful than our past 100,000 years? Through SETI, I suggest that it would be well justified to seek the wisdom of far more advanced societies from our common universe. The worst case result would be no signal, but a lot of useful science, at a cost far less than the AIG bailout.

Comments (7): One respondee was particularly persistent. He believed in flying saucers and felt that our government was hiding the evidence. I think he's seen too many movies and ...well, let me stop there. I had this to say:

Well, you certainly are entitled to your opinion, but my personal belief is that flying saucers are not real. Yes, sometime over the past couple of billion years some alien form might have visited us, and perhaps even seeded life, but the most scientists don't give any credence to all that you mention above. Especially those who work on the concept of Search for Extraterrestrial Intelligence. On the other hand, I don't believe in an afterlife, but 90% of Americans do.

*The following is a non-HuffPo. For a reason I couldn't understand, they passed on publishing it. Maybe it was too similar to the previous essay. Anyway, this piece is out of chronological sequence, for it was written in June of 2010, but I insert it here to follow my other ET essay. In the **Table of Contents**, these non-Huffpos will have capped titles.*

Seti: Part One

The **Extraterrestrial** (ET) was a movie, and the concept of such intelligence (ETI) has challenged our best minds from the beginning of humanity. The early Greeks showed progressiveness, and Metrodorus of Chios[178] in the fourth century BC wrote in his **On Nature**:

To suppose that Earth is the only populated world in infinite space is as absurd as to believe that in an entire field sown with millet, only one grain will grow.

In November of last year I began to serialize Chapter 4 on Search for Extraterrestrial Intelligence (SETI)[179] from **SOLUTIONS for Humanity**. To gain an appreciation of distance, it is useful to link this dimension with time. A light year is not a dimension of time. This is the distance light travels in one year. Rounding it off, call it 6 trillion miles (6,000 billion miles or 6,000,000 million miles).

Thus, one light year is a very long distance. To travel 6 trillion miles, you would need to make more than 30,000 round trips from Earth to the Sun.

What inspired me to write this piece was a recent report indicating that Kilauea Volcano, the subject of my most recent posting,[180] has now continuously erupted for 10,000 days. Eyjafjallajoekull, that Icelandic volcano which caused havoc with European air travel, while apparently now subsiding, has only been active for about a month. I wondered, anyway, how long, really, was a timeframe of 10,000 days in the cosmic perspective of time.

I was on the Big Island of Hawaii for the New Year period in 1983. On January 3, I happened to be golfing at the Volcano Golf and Country Club. I forget exactly at what hole, but in the back nine, the ground shook. We then noticed that a couple of miles away a lava fountain appeared at Puu Oo. Kilauea Volcano has now been continuously erupting for 10,001 days. This is close to a billion seconds. In comparison, Jesus Christ (if he existed) was born about a billion minutes ago.

I had returned to the University of Hawaii a few months previously after a three year assignment working for the U.S. Senate. Much of my professional life thusly unfolded, plus

---

178 http://www.roger-pearse.com/weblog/?p=3894

179 http://planetearthandhumanity.blogspot.com

180 http://www.huffingtonpost.com/patrick-takahashi/volcanoes-hawaii-versus-i_b_543571.html

a decade of retirement...and that was about a billion seconds ago. Light traveled 186 trillion miles during this 27+ year period.

The closest star is Proxima Centauri at 4.2 light years, so if intelligent life lived there, we could have have sent three messages to each other during this period. But our Milky Way Galaxy is huge. Light would take 100,000 years from one end to the other. Another way of contemplating all this is that during this period from when I saw that eruption of Kilauea Volcano in 1983 till today, light would have travelled just 2.7% across just our galaxy, and ours is not a particularly large one.

Thus, how vast is space? Best estimates are that there are from 200 billion to 400 billion galaxies[181] in our universe, and each galaxy has, oh, 100 billion stars. Interesting, though, that our neighboring galaxy, Andromeda, is closer to Planet Earth than the center of our galaxy. Thus, it would take light (or a really strong electronic signal) all of 25,000 years to reach the edge of the Andromeda Galaxy from Hawaii. There are possible parallel universes and such, but let's stick only with ours, for now.

So, if you're wondering if flying saucers regularly visit us from other stars--and HuffPo does publish these postings[182]--consider that the fastest man-made object is Helios 2 at slightly more than 150,000 miles per hour. A spacecraft travelling at that speed would take almost 20,000 years to reach Proxima Centauri. If you can remember that far back, Pioneer F(also known as #10, the one with info about our civilization) is headed for Aldebaran, a star 65 light years away. We long lost contact with this probe, but if it succeeds in getting there, you will need to wait another 2 million years.

But why go to Aldebaran. Let's shoot for Proxima Centauri (which, unfortunately, is a Red Dwarf, and pretty useless for life, but we'd be in the neighborhood of Alpha Centuari, only a few miles further) and not be limited by current knowledge. Nuclear pulse propulsion (which was an earlier Project Orion,[183] cancelled by the 1963 test ban treaty), should have been able to get us there in 85 years. But, for what? More so than the relevance, it would probably take 100 times the current annual energy use of the entire planet.[184]

Today, I recommend that we refine the astroscience, but not spend anything on actual manned space travel, kind of what President Obama proposed for the 2011 budget,[185] which irritated a few astronauts. We landed on the Moon, and that was worth the ultimate fracturing of the Soviet economy. But international grandstanding was apparently necessarily to end the Cold War. Today, we have no equivalent enemy worthy of such grandeur. Yes, let's advance

---

181 http://www.universetoday.com/guide-to-space/galaxies/how-many-galaxies-in-the-universe/

182 http://www.huffingtonpost.com/patrick-takahashi/volcanoes-hawaii-versus-i_b_543571.html

183 http://www.ted.com/talks/george_dyson_on_project_orion.html

184 http://futurismic.com/2008/08/20/rocket-scientists-reaching-the-nearest-star-in-a-human-lifetime-is-nearly-impossible

185 http://www.compositesworld.com/news/proposed-2011-nasa-budget-cuts-moonmars-program-favors-commercial-orbital-scheme

knowledge, but the $120 billion Bush the Younger plan and $500 billion Bush the Elder proposal for Man on Mars can be delayed...by about a millennium.

There are some, Erich von Daniken,[186] for example, who argue that ET's already were here, and, in fact, might well have been those gods that developed our various religions. The more rational ignores this point of view, and suggests that economic and technological conditions today are such that, while we can't afford to go anywhere, we should at least search for extraterrestrial intelligence (SETI). Congress explicitly has forbidden NASA to spend money on SETI. Part 2 will build from my time with the NASA Ames Research Center on detecting extrasolar planets to where the field of SETI is in the Year 2010, already almost a decade past Kubrick's *2001*.

Comments: There were no comments because this was never published before. Part 2, effectively can be the previous essay. The field has not changed much.

---

186 http://www.daniken.com/e/index.html

On 31March09 I jumped in on the economic crisis again, and harped, again, on how Detroit could take control of its destiny. I was also concerned that the Obama Administration had the wrong priorities in ground transport.

A Solution for the American Auto Industry[187]

President Barack Obama took steps to re-invent General Motors and Chrysler for the better, hopefully. The current focus is on survival. Unaddressed was how to regain American leadership in the automobile industry.

President Obama earlier this month visited the Edison Electric Vehicle Center located in Pomona, California, where he spoke glowingly of this option and mentioned the availability of large development funds. His stimulus plan, he said, provides $2.4 billion to help Detroit make the transition to hybrid cars. He also talked about a $2 billion battery R&D program to compete with the world, which could well be the same program.

A year ago I would have rejoiced at his dedication to EVs. Today, I'm not quite so sure. I've written two posts on this related subject, comparing plug-ins with the fuel cell and suggesting that we develop our own technology for powering cars.[188]

The vaunted GM plug-in Volt, for example, will sell beginning in 2010 for $40,000. As an intermediate step, I guess this is the best they can do, but this is no way to reassert American dominance in the field, for the Volt will use a lithium battery from South Korea. Why? It appears that countries from Europe and the Orient have a lock on workable next generation battery patents for lithium. The sad conclusion is that we have no future in battery powered vehicles.

The announced American battery consortium[189] of 14 companies with the Argonne National Laboratory (note, particularly, the absence of GM, Ford and Chrysler) is seeking billions to advance our cause. Their announced focus will be on lithium, so one immediately can speculate that all they will be able to do will be to streamline the marketing and importation of these batteries. Why don't they, instead, produce a better battery, you say? Well, it turns out that battery technology has reached the end of the line. There is no future material on the horizon, save for maybe some mysterious super capacitor or blue-sky nanotech pathway. Mind you, batteries will still be a part of a fuel cell hybrid system, so do improve the ultimate battery.

---

187 http://www.huffingtonpost.com/patrick-takahashi/a-solution-for-the-americ_b_180777.html

188 http://www.huffingtonpost.com/patrick-takahashi/is-there-an-option-more-p_b_150824.html

189 http://seekingalpha.com/article/111466-u-s-battery-consortium-seeks-1b-is-it-a-waste

Why then don't we use some American ingenuity to develop a superior way for moving vehicles? Immediately scratch the ICE and lead acid batteries. Heartbreak, but eliminate the nickel hydride battery, an invention of American Stanford Ovshinsky of Detroit. His travails with GM deserve a tragic re-write by the next Shakespeare. Also delete the lithium battery, for we missed the boat here. What else is there?

There is a technology that was invented 170 years ago in Wales. It is called a fuel cell and is used on NASA journeys to produce electricity and freshwater for drinking. Being readied for commercialization is the micro fuel cell to run your iPod and portable computer. For the same space as a lithium battery, this device can operate five times longer. The fuel is methanol.

Methanol is the simplest alcohol and can be directly fed into a fuel cell without reforming, a very expensive process. Now this is difficult to believe, but one gallon of methanol has 1.4 times more accessible hydrogen than one gallon of liquid hydrogen.

Today, methanol is produced through the steam reforming of natural gas, but in the future, biomass can be gasified and catalyzed into biomethanol. This is a natural for the farm industry, for all the non-edible portion of any crop can be collected for processing into methanol.

Why then don't we do this already? Well, it turns out that the Farm Lobby a decade ago came up with what then was a brilliant idea. Why don't we ferment corn into ethanol, which can be used to reduce oil imports. Then, the price of corn will rise and farmers will be more successful. But prohibit methanol, for it is too cheap and will affect the marketing of ethanol. The ploy worked!

Now that people are beginning to wise up to using food for fuel, as the laws are already in place, the Farm Lobby is turning to cellulosic ethanol. While this is technically possible, it will be an economic disaster.[190] If you have any relatively dry biomass, it is much cheaper to produce methanol. As this methanol is the perfect fuel for the direct methanol fuel cell (DMFC) to power cars, what a match made in heaven between the heartlands of American and Detroit.

But we have a problem. There is no DMFC for cars. Our Department of Energy has purposefully banned any methanol R&D. Details on parameters such as politics, safety, energy density, etc., can be found **in *Simple Solutions for Planet Earth***.

<u>This is then the golden opportunity for our new administration seeking change</u>. As other countries usually watch what we're doing, they also have not done much in this area. The Japanese are on the cusp of commercializing the DMFC for portable applications, but a device for vehicles is a decade away.

Rather than spending billions on a next generation battery to nowhere, we have a once in a lifetime chance to take a leadership role in the solution to Peak Oil, Global Warming and Economic Development. We need a Marshall (but call it Obama) Plan for Detroit to develop

---

190 http://www.grist.org/article/cellulosic-ethanol-not-likely-to-be-viable/

the Direct Methanol Fuel Cell for vehicles. In time, we will be able to export products instead of burying ourselves further by quickly converting to plug-in vehicles where the batteries will need to be imported.

Comments (13): There was a healthy discussion on methanol as a fuel for ground transport. There was universal agreement that this made sense.

*How quickly a year goes by. My 13May09 posting summarized my 52 articles. I must say, this was a productive period: an article a week for **HuffPo**, a blog a day[191] and two books, **SIMPLE SOLUTIONS for Planet Earth** and **SIMPLE SOLUTIONS for Humanity**. I also do other things…like golf.*

My Year With The Huffington Post[192]

My very first posting, on May 29th 2008, was written during the heat of the Obama-Clinton race for the Democratic nomination, and was entitled "Well, Barack, We have a Problem."[193] I suggested that he was the only individual capable of ameliorating the economic cataclysm to come. Well, two for two, as he did become President and seems to be making the right moves.

I further intimated that the best way to secure funds would be to reduce the defense budget. He did, sort of, but nothing close to the 10% cut I recommended, to be followed by 10% each year if all other countries did the same. Soon, then, his legacy would be: Peace on Earth forever. Ridiculous, perhaps, but something to consider.

The thought of posting an article a week for the entire year was way beyond my wildest dreams, but here I am, in mid-May, with this, my 52nd article. A couple got no comments, but one went beyond a hundred.

In general, as traditional columnists report on the obvious, I thought I would focus on the unobvious. Yes, except for the Farm Lobby, we all know that ethanol is bad,[194] but my answer is the direct methanol fuel cell. There was very little support, especially none from the U.S. Department of Energy. I even went so far as to express caution regarding the plug-in electric car[195] and the hydrogen economy.[196] I got heartburn from the responses. Just wait, though, for I predict that the biomethanol economy will be just around the corner (okay, at least ten to twenty years away).

No one writes about ocean opportunities. Former President George W. Bush, for example, took pride in closing down portions of our ecosystem to development and the National Oceanic and Atmospheric Administration has a preservationist mentality. A former chief scientist waxed about her fishy friends and their only ocean technologist was fired more than

---

191 http://planetearthandhumanity.blogspot.com

192 http://www.huffingtonpost.com/patrick-takahashi/my-year-with-the-huffingt_b_202673.html

193 http://www.huffingtonpost.com/patrick-takahashi/well-barack-we-have-a-pro_b_104201.html

194 http://www.huffingtonpost.com/patrick-takahashi/ethanol-vs-methanol_b_106380.html

195 http://www.huffingtonpost.com/patrick-takahashi/is-there-an-option-more-p_b_150824.html

196 http://www.huffingtonpost.com/patrick-takahashi/simple-solutions-for-our_b_146906.html

decade ago. As NOAA reports to the Department of Commerce, you would think that at least one office would promote intelligent commercial development of the seas. Nope.

I cheer Patri Friedman and his Seasteading, Inc., and provide encouragement to Ted Johnson and his OTEC crew from Lockheed Martin and Makai Ocean Engineering, but there seems to be no real interest in something I have termed, the Blue Revolution,[197] to develop the seas for humanity in harmony with the marine environment. There is almost a unanimity of world opinion that we screwed up our lands and atmosphere, so, let's not destroy our ocean. Nonsense, for sustainable resources, green materials, exciting new habitats and more can become our future if we do it right. And improve the environment in the process.

One of my recent tactics is the use of doomsday, for if sound logic does not work, maybe fear might. I have a few friends who are preparing for the end, and even an active internet forum on the general subject. *My Daily Blog*[198] covers these activities, including serializing those two books on **Simple Solutions**. As of today, 73 countries have visited that site.

Anyway, one of those chapters is entitled The Venus Syndrome,[199] hinting that the vast deposit (said to be twice as much more than the known amount of oil, coal and natural gas) of marine methane hydrates at the sea bottom could well reach the surface and turn Planet Earth into Planet Venus at 900 degrees F. Methane, as most don't know, compared to carbon dioxide, is more than 20 times worse for inducing global warming. A novel on the subject is in the works.

I carped on why we have no national energy policy[200] (this most prosaic of postings drew more than 100 comments), why Republicans like fossil fuels and not care that much for the environment,[201] the amount we spend on national security,[202] and our more recent tendency to copy rather than build something better.[203] On the more positive front, I explained why renewable electricity is our only viable option,[204] identified the ideal biofuel for development[205] and provided a solution for our domestic auto industry[206]. You will be stunned about how Americans view "Evolution, Global Warming, Doomday and the Afterlife"[207] and might still be confused about the ethnicity of Barack Obama.[208]

---

197 http://www.huffingtonpost.com/patrick-takahashi/blue-revolution_b_166977.html

198 http://planetearthandhumanity.blogspot.com

199 http://www.huffingtonpost.com/patrick-takahashi/the-venus-syndrome-part-o_b_106120.html

200 http://www.huffingtonpost.com/patrick-takahashi/why-is-there-no-national_b_104507.html

201 http://www.huffingtonpost.com/patrick-takahashi/why-do-republicans-like-f_b_108190.html

202 http://www.huffingtonpost.com/patrick-takahashi/why-do-we-spend-so-much-o_b_116535.html

203 http://www.huffingtonpost.com/patrick-takahashi/america-dont-copybuild-so_b_166134.html

204 http://www.huffingtonpost.com/patrick-takahashi/renewable-electricity-is_b_162435.html

205 http://www.huffingtonpost.com/patrick-takahashi/simple-solutions-for-our_b_146906.html

206 http://www.huffingtonpost.com/patrick-takahashi/a-solution-for-the-americ_b_180777.html

207 http://www.huffingtonpost.com/patrick-takahashi/evolution-global-warming_b_168827.html

208 http://www.huffingtonpost.com/patrick-takahashi/barack-obama-is-gray_b_133101.html

As miserable as the world has been over the past year, we now have no Cold War and will soon exit Iraq, do not worry much about the population bomb and, by the way, what happened to acid rain? The economy is now rising, there is change in D.C. and the Sun will continue to shine.[209] Thank you for a wonderful year with *HuffPo*.

Comments (11): *This was my most heartwarming set of responses. Friends and the general readership actually praised me for my efforts. What shocked me, though, is that I now realized I wrote an article a week for the* **Huffington Post**. *And all for free. One more thing. I said that 73 countries had visited my blog site. Today as I write this (18June2010) review, the number is up to 149.*

---

209 http://www.huffingtonpost.com/patrick-takahashi/the-sun-will-continue-to_b_155160.html

I experienced the Avian Flu hype from the late 1990's into the early 2000's. The watchers are scared beyond any reality about any strain morphing to both be more contagious and with higher mortality. The 1918 flu was an epidemic, but that was the only exception. When the Swine Flu hit, my first reaction was, oh, oh, another overreaction coming. The difference is that people are overly concerned about their personal and family life, like an epidemic or airport security. We are not so troubled by things like Peak Oil and Global Warming. Thus, on 3June09 I just had to comment on the true pandemic: the media hype.

A Pandemic Worse than the Swine Flu[210]

Five billion are killed by a virus. Sound familiar? Think the reaction to swine flu. The 1995 science fiction film starring Bruce Willis and Brad Pitt called *12 MONKEYS*[211] identifies a terror group implicated in this plot. This was a red herring (diverted attention).

Willis failed in his return from the future and doomsday was not prevented. It occurred to me that our current imagined pandemic might well be the modern day equivalent. Again, we are focused on the wrong target. Maybe there is a lesson to be learned.

As a starter, let us review where we are on this subject. RJ Eskow provided in HuffPo an excellent summary, entitled, The Meaning of Swine Flu, the Universe, and Everything.[212] If you read through the comments and trace some of the references, you can take a comedic pathway leading to Jeff Horwich's Don't Cough on Me Alejandro (sung to Don't Cry for me Argentina),[213] a satire found in Face Book. Also, too, you just can't skip another HuffPo posting, this one in the Comedy section, by Juliet Jeske, on Congresswoman Michele Bachmann's Guide to the Swine Flu.[214] You will then be trapped into yet another Comedy HuffPo, this one by Will Menaker,[215] on CongressBorat Bachmann, from Minnesota, always in the top ten among states in educational achievement and well being (health). Yet, funny, but they produce a Governor Jesse Ventura and can't extricate themselves from the continuing farce with Al Franken, known as their senatorial race. But I digress.

You might think that I am making a joke of the swine flu, for my April 24 posting was entitled, Benefits of the Swine Flu Scare,[216] followed on May 5 with Cinco de Mayo and the Swine Flu.[217]

---

210 http://www.huffingtonpost.com/patrick-takahashi/a-pandemic-worse-than-the_b_207226.html

211 http://www.imdb.com/title/tt0114746/

212 http://www.huffingtonpost.com/rj-eskow/the-meaning-of-swine-flu_b_205793.html

213 http://www.facebook.com/video/video.php?v=85559888622&ref=nf

214 http://www.huffingtonpost.com/juliet-jeske/michele-bachmanns-guide-t_b_195901.html

215 http://www.huffingtonpost.com/will-menaker/michele-bachmann-revealed_b_198054.html

216 http://planetearthandhumanity.blogspot.com/2009/04/benefits-of-swine-flu-scare.html

217 http://planetearthandhumanity.blogspot.com/2009/05/cinco-de-mayo-and-swine-flu.html

But no, my message is deeper, for these seeds of hysteria provide clues about our future. You can wonder about how we got ourselves into this dilemmic mode: on the one hand, we have something that will almost never happen (a serious swine flu epidemic), while on the other, there is death, worried mothers and panic.

Let's first look at the reality. You can go to my Chapter 2 of *SIMPLE SOLUTIONS for Humanity,* or *Planet Earth and Humanity,* where hard numbers are provided. To summarize:

1. The regular flu kills one in a thousand, while the swine flu seems closer to one in a hundred. That previous avian flu of a decade ago has a 60% mortality rate, but it is not all that contagious. The fear is that the swine flu will morph into a more dangerous form. The potential of this happening is very low, and, in any case, it is appearing that all 50 or so variations[218] might turn out to be treatable with one vaccine. So, certainly, continue the vigilance and spend my tax dollars to develop a common vaccine, but don't close down countries and schools, and, by the way, tourists can safely return to Mexico and Hawaii.

2. The numbers are embarrassingly obvious. Since the so-called epidemic was announced only a little more than a month ago, we have had about 500 cases and 3 deaths a day. I wouldn't want to be one of those statistics, but consider that a million people daily contract some form of flu and at least a thousand die, every day, usually from complications (heart, pneumonia, etc.). This terrifying swine flu is thus hardly detectable noise. I might further add that traffic fatalities number 3000/day, but we drive on.

Why, then, has the world, epitomized by the World Health Organization (which can best be appreciated if you know the internal workings of the United Nations), gone bonkers over the swine flu? I would like to speculate on the reason. I think it has to do with our political way of life influenced by the world wide web (WWW), as sensitized by the terroristic act of September 11, 2001. Add the palpable need to cover your rear.

First came airport security. From all reports, the vast funds allocated and the time we waste have not made us any safer.[219] Yet, the great majority of the public likes it.[220] Of course this is symptomatic of our world today where most believe in the afterlife but have doubts about evolution and global warming.

In particular, the masses are supersensitized to any threat. So when you mix in the World Wide Media (WWM) and WWW, the reaction can be instantaneous and overwhelming. I early recognized the power of this medium, so thought that the *Huffington Post* would be an ideal vehicle to share my thoughts, for the instant feedback feature of these virtual portals, in my mind, would provide power to the people, replacing protest marches. Clearly, I have not learned how to galvanize action because my 52 *HuffPostings* have influenced few, if any.

---

218 http://articles.latimes.com/2009/may/23/science/sci-swine-flu23

219 http://www.wired.com/wired/archive/12.03/view.html?pg=2

220 http://www.usatoday.com/travel/flights/2008-12-08-airsecurity_N.htm

Part of this failure I ascribe to the fact that I tend to focus on Peak Oil, Global Warming and related topics. There is no threat of early death in my articles.

Swine flu, though, conjures dark images of your mortality. The communications industry, like CNN, saturates air time on such issues because they know people will watch. The WWW picks it up and decision-makers are hopelessly influenced. The cascading circle of information gains a life of its own. **The truth is that the truly dangerous virus is not the swine flu, but the medium itself. The pandemic is this resultant overreaction.**

My simple solution is to jump on the bandwagon and complete ***THE VENUS SYNDROME: The Novel.*** Taking a cue from this posting, if compelling logic does not work for global warming, perhaps fear might.

Comments (10): *There were several decent comments. However, we never did come anywhere close to immunizing the world wide media from continuing to over hype what has later turned out to be another empty scare.*

My first year with the Huffington Post was mostly centered on scientific subjects. In Year Two, I began to treat subjects from SIMPLE SOLUTIONS for Humanity. Chapter 1 starts with crime. So I began with a solution to crime on 11June09.

A Simple Solution for Crime: Background[221]

Purpose

I've been wondering why some of my postings draw no comments, and a rather tedious one, "Why is there No National Energy Policy?"[222] attracts more than a hundred. My first year with the Huffington Post focused on energy, the environment and the economy, more or less the subject of my first book, *Simple Solutions for Planet Earth* (*Book 1*).

My second book, *Simple Solutions for Humanity* (*Book 2*), wanders into crime, religion, eternal life, the search for extraterrestrial intelligence and the more sociological areas. I'm curious whether these subjects can entice a few more to comment, not for the sake of a response, but to catalyze an effort to advance our society.

So, I'll start with what might well be the most controversial of all, a simple solution to crime. Remember, we place more of our citizens in jail than any other country, and are now faced with what to do with Guantanamo. There is an opportunity here to institute significant change from the current non-functional norm.

Background

Draco wrote the first constitution of Athens. His laws were severe, as a debtor became a slave and there was death for even minor offenses. In time, this form of punishment became known as Draconian. The solution to crime might well suit this term, and is from Chapter 1 of *Book 2*.

There is almost universal agreement that the best way to prevent crime is through careful upbringing and the education process. Every care and any reasonable expense should be expended to mold a solid citizen. Unfortunately, there will still be crimes.

Civilization has exacted a range of punishments over time. In Exodus 21:23-27, an "eye for an eye" provided the principle of proportionate punishment. Equitable retaliation was the belief. This same concept is presented in the Code of Hammurabi. Of benefit to the criminal, this type of law prevents excessive punishment. The Quran actually urges the victim to accept a

---

221 http://www.huffingtonpost.com/patrick-takahashi/a-simple-solution-for-cri_b_213530.html

222 http://www.huffingtonpost.com/patrick-takahashi/why-is-there-no-national_b_104507.html

lower compensation, if not totally forgive, with an unwritten promise of a later reward, such as, perhaps, paradise. Martin Luther King has been quoted to say, "the old law of an eye for an eye leaves everyone blind."

Fair enough, as everyone makes a mistake now and then. If the crime can be forgiven, fine, now do everything possible to remediate that person. Be protective, be reformative. Rehabilitate. Much more can be done here.

Then, for some, a second crime is committed and there is another conviction. Now what? Is there a way to prevent habitual offenses? Do longer and longer prison sentences make sense? Maybe, but the cost factor troubles me.

Alas there is that third strike. Chaperoned, air-conditioned comfort, for life?

In baseball, if you get three strikes, you are out. In parts of the U.S., with three strikes, or convictions, you are in, jail, that is, maybe even for a long time, perhaps life. This is a relatively new concept, which started in the state of Washington in 1993 and California the following year, with three felonies and you're in prison for 25 years to life. Their prisons are now overbooked by more than a factor of two.

Kevin Weber was sentenced to 26 years to life for stealing four chocolate chip cookies (actually, that was only the evidence from a robbery). There are hundreds of shoplifters in prison for 25 years or more, one who stole a $3 magazine. Les Miserables? Many of our prisons now hold double the designed capacity, and more. And, you're paying for all this. Is there a better way?

More than half of the states now have some form of Three Strikes and You're Out law. In March 5, 2003, the U.S. Supreme Court ruled that this was not "cruel and unusual punishment."

According to a 2005 report of the National Crime Victims' Rights Week, crime costs Americans somewhere in the range of $500 billion annually. However, in 1999, David Anderson[223] published in the ***Journal of Law and Economics*** a paper outlining that the net burden of crime in the U.S. was $1.7 trillion per year, or $5667 for each person in the country. This did not include the physical pain, long-term mental anguish and general fear anyone has about stepping out at night and such. Wouldn't it be wonderful if we could just eliminate this compromise in our daily life forever? The savings of more than $2 trillion/year would be wonderful, but just the lifestyle enhancement alone would be huge.

Here is my problem with Three Strikes and You're In Jail for a Long Time. As legally necessary as keeping someone in jail for the rest of his life might be, I think it is not worth my tax dollars. Does this mean I am against the concept of keeping criminals away from me? No, I like this, only, let us be truly sensible, maybe, even draconian. First, do we want a better educational system or should we build more prisons? Second, isn't it cruel, anyway, to keep

---

223 http://www2.davidson.edu/news/news_archives/archives99/9910_anderson.html

someone locked up for a lifetime? Third, how, really, can we actually best prevent that fourth crime.

Regarding the economics, there are studies that show it is cheaper to keep someone in jail for long periods than the judiciary cost of capital punishment. Then, there are other studies that disagree. I call for improving the efficiency of that third and terminal judgment.

While the California Three (felony) Strikes and You're Out law is already controversial, in Part 2 to come of this series, found in the 13June09 issue[224] of my daily blog, I offer a Three (any crime) Strikes and You're Dead concept for consideration as an ultimate solution for crime.

*Comments (6): I expected a flurry of irate responses to my concept of Three Strikes and You're Dead. But this was supposedly coming in Part 2, so maybe they were waiting. Alas, the **Huffington Post** refused to publish this follow-on article. I guess they do have limits, for this was the first one they had refused to print. Following is the version they declined.*

---

224 http://planetearthandhumanity.blogspot.com/2009/06/simple-solutions-for-crime-three.html

*Well, here it is, the posting that **the Huffington Post** refused to print!*

A Simple Solution to Crime:
Three Strikes and You're Dead

Let me come straight to this simple solution to crime: **Three Strikes and You're Dead (TSAYD)** To repeat, **THREE STRIKES AND YOU'RE DEAD!!!**

First conviction: Someone commits a crime. Do everything possible to reform the misguided individual. One remediative option is restorative justice, promoting repair, reconciliation and the rebuilding of relationships. The process attempts to build partnerships, seeking balanced approaches for victim, wrongdoer and community. There will be a lot of counseling and, as necessary and possible, some restitution.

New Zealand legislated in 1989 restorative justice for juvenile crime. Then, there were 64 violations per 1,000 in the population. Today, this figure has dropped to 16.

The concept is hardly new, as the **Pentateuch**,[225] the so-called books of Moses, advocated compensation for property crimes, as did the Code of Hammurabi (1700 BC). The Code of Ur-Nammu[226] (2000 BC) required amends for violence.

Spend the bigger bucks on shoring up the early life (as underscored in Part 1), but be generous, too, on reclaiming this individual after a tolerable first offense. The primary objective will be to prevent those potential second and third crimes. Half of all inmates are back in jail only two years later. A simple way to reduce this rate is to strike fear into the offender so that he abandons all thought of committing another crime. Thus, the penalty for entering stage two should be terrible, if not horrendous.

Crime #2 is committed and the defendant is judged guilty. Standard prison? Nope. Save your tax monies to build better school systems. Find some hellacious environment where the prisoner will need to support himself, and where the cost to society will be minimal. A mild form -- I was thinking more in terms of dungeons or caves -- of this concept is represented by Joe Arpaio, sheriff of Maricopa Arizona County:

- jail meals, such as bologna sandwich, cost 40 cents/serving, and he charges inmates for them;

- no smoking, no coffee and no porno magazines;

---

225 http://bible.org/article/introduction-pentateuch

226 http://www.realhistoryww.com/world_history/ancient/Misc/Sumer/ur_nammu_law.htm

- chain gangs to do free work on county projects;

- took away cable TV, but was forced to put them back because a federal court required that for jails, so he played only the Disney and weather channels, and added Newt Gingrich lectures;

When inmates complained, he told them don't come back. Two thousand prisoners in tents with no air conditioning, even when the temperature is more than 116 degrees F. Sheriff Joe says that our soldiers in Iraq are in tents where the temperature exceeds 120, in full battle gear.

Ah, but Sheriff Joe is in trouble with the Lord. A Catholic scholar wants communion withheld from him.[227]

The question is, will those who survive be more apt not to return? The answer is, heck, yes, but mostly because of the consequences of a third conviction! Anyway, why waste good money on the hopeless, which is defined as anyone who is stupid enough to commit a second crime knowing that the punishment will be hell. The prison of Sheriff Joe is a reasonable simple solution for strike two. There are harsher, and probably more effective (but illegal and immoral) options, but for now, let us leave them for future consideration.

Third conviction: termination! Yes, death. The U.S. has now had more than 1,000 executions since the Supreme Court ended a moratorium three decades ago. The transition will be messy, but under the TSAYD formula, this number could seriously increase in the first few years, but should decline with time. The odds are astronomically high that crime rates will significantly drop within the decade. Where are the supporting statistics? Probably none exists. I just feel this way.

Oh yes, there are millions of questions and issues. What about a white collar criminal or traffic violator? Can you execute someone for stealing a chocolate chip cookie or a magazine? You've no doubt read of a car thief, who is convicted, convicted mind you, dozens of times (well, I'm not sure what the record is, and I can't imagine our court system being so efficient as to actually convict the same person so many times), and still somehow runs loose and is arrested for stealing yet another car. Maybe enough should be three convictions for anything. Under the TSAYD system, the odds are high that this will all stop after the first arrest.

But what about those dependent on drugs who cannot control themselves or the few so idiotic as to commit that third crime? Answering this question will only irritate churches, social workers, the American Civil Liberties Union and mothers.

Then, there is that matter of mental illness. A Bureau of Justice Statistics study reported in 2006 that half of those in jail suffered from mental health problems, especially those convicted

---

227 http://www.examiner.com/x-35821-Immigration-Reform-Examiner~y2010m6d18-Catholic-scholar-wants-communion-withheld-from-Sheriff-Joe-Arpaio

three or more times. Sounds menacingly Hitleresque, but perhaps an argument can be made that the elimination of this drag on society would only improve the quality of life for the productive and innocent. Former Education Secretary William Bennett,[228] on his radio show, is quoted to have said:

To reduce crime, you could -- if that were your sole purpose -- you could abort every black baby in this country, and your crime rate would go down...That would be an impossible, ridiculous and morally reprehensible thing to do, but your crime rate would go down.

He said his statement was taken out of context, and TSAYD definitely falls in this category.

Could a "Three Strikes and You're Dead" law ever be enacted anywhere in the world today? I hope not. But let me suggest that the concept be debated and considered, for, perhaps, a compromising intermediary first step might well be a more universal application of the Sheriff Joe penal system. However, after considerable public anguish, I would not be surprised if something a lot more draconian, but justified, actually is attempted. What somewhat worries me (because I am compassionate, honestly), though, is that many of my friends in casual conversation, with few exceptions, actually like the notion of TSAYD.

Thus, the nature of our present culture, maybe even epitomized by the liberal readership of the **Huffington Post**, could well be the primary impediment to ending crime forever. This could well be the legacy of our imperfect, but reasonably mature, society.

---

228 http://blogcritics.org/politics/article/bill-bennett-exterminate-blacks/

On 29June09 I was looking ahead to the upcoming Kyoto II in Copenhagen. I had a nightmare that the U.S. was going to unleash cap and trade on the world, even though, of course, Europe has been choking on it for some time now. Well, the worry was unnecessary as nothing happened, neither congressional action nor summit agreement.

The Carbon Dioxide Credit Program[229]

We all know now that the House of Representatives barely passed a global warming remediation bill last week, known as the American Clean Energy and Security Act of 2009. I should be rejoicing, for compared to the Bush Administration, the Obama White House has been remarkably progressive about renewable energy and becoming a responsible partner to check global warming. Yet, I'm not, and equally concerned about this legislation are the Nature Conservancy and Friends of the Earth.[230] Of course, many Republicans are apoplectic, but that's expected.

There are various **HuffPos** commenting on this victory for President Obama, and I can cite two in particular that provided useful information: "The Simple Arithmetic of Global Warming:[231] and "Cap and Trade Explained."[232] These two postings supported the legislation. I would like to provide a different look for the sake of longer term implications.

There are two basic approaches to reduce fossil fuel consumption. Cap and trade is the version today popular with politicians. The second is the carbon tax. Anyone with any knowledge of this subject will tell you that the carbon tax is more sensible than cap and trade. So why is cap and trade being foisted on the public?

This is simply because the carbon tax sounds like a tax. Republicans are inherently opposed to more taxes. As both houses of Congress are so dominantly Democrat, with a Democrat in the White House, what is, then, the big deal, for Democrats who are not totally allergic to taxes?

Furthermore, this is oh so ironic because people don't directly pay this tax. Companies that burn coal to generate electricity or refine crude oil into gasoline are so burdened. Then, of course, because these fuel suppliers need to make a profit, they add this charge to your electricity and gasoline bill. This then becomes an indirect tax to the masses.

But there should be general acceptance by the people if much of this extra charge is then credited to your account when you file your taxes. Yes, of course, the cost of everything then

---

229 http://www.huffingtonpost.com/patrick-takahashi/the-carbon-dioxide-credit_b_221933.html

230 http://www.thedailygreen.com/environmental-news/latest/house-climate-bill-47062602

231 http://www.huffingtonpost.com/david-fiderer/the-simple-arithmetic-of_b_221689.html

232 http://www.huffingtonpost.com/2009/06/25/the-climate-bill-explaine_n_221233.html

also goes up, but all polls show that the American public is supportive of the Renewable Electricity Standard.[233]

But how much more would we be willing to pay? A Financial Times/Harris Poll[234] of early last year showed that both Europeans and Americans do not want to pay any more for renewable electricity and clean gasoline substitutes. Others say that the scare of $147/barrel oil a year ago has changed the attitude of most of us. But how much will electricity and gasoline increase in cost if a meaningful carbon credit (let's not call it a tax) is instituted?

A 5 cents/pound carbon dioxide credit on petroleum would translate to $1/gallon more at the pump, or the $3/gallon some of you currently pay would go up to $4/gallon, exactly where we were in July of last year when oil prices peaked. Coal fired electricity would increase by about 10 cents/kWh. As the current price is about 11 cents/kWh, that would almost double the price to 21 cents/kWh. In comparison, my **HuffPo** of earlier this year[235] reported that the California Public Utilities Commission indicated that a coal fired power plant with carbon capture would be expected to generate electricity at 17 cents/kWh. Rightfully, these facilities should not pay this assessment. In that same article Joseph Romm seems to feel that nuclear power should not be a factor, for the cost of electricity from NEW nuclear facilities would fall in the 25 cents to 30 cents/kWh range.

Would a 5 cents/pound carbon dioxide credit make a sufficient difference to ameliorate global warming? Well, the March issue of **Scientific American**[236] indicated that, while wind and geothermal power could be produced for 7 cents/kWh, solar thermal electricity still costs more than 20 cents/kWh and solar photovoltaics are at almost triple that rate. However, there is reasonable hope that those next generation thin film PV systems will be much cheaper.

Thus, the swift enactment of a 5 cents/pound carbon dioxide credit would be very effective, indeed. But is there any way to just toss away the term carbon tax and replace it with the carbon dioxide credit program, where 100% of the credit collected by the government would be returned to the taxpayer? Well, the Senate has not yet acted. Plus, the general reaction to the House measure is that this legislation is seriously flawed because of cap and trade. Unfortunately, the Senate version will be cap and trade, so, unless there can be a dramatic readjustment, I'm afraid our politically compromised global climate warming remediation package will become the centerpiece for the World to embrace when the successor to the Kyoto Protocol is discussed this coming December in Copenhagen. That very much concerns me.

---

233 http://tdworld.com/substations/awea-renewable-standard-poll-0509/

234 http://eponline.com/articles/2008/02/29/poll-people-dont-want-to-pay-more-for-renewable-energy.aspx

235 http://www.huffingtonpost.com/patrick-takahashi/renewable-electricity-is_b_162435.html

236 http://planetearthandhumanity.blogspot.com/2009/06/on-twitter-and-cost-of-renewable.html

<u>*Comments (12)*</u>: *Tom Burnett was particularly vocal, and his arguments made sense. Madeleine Austin was also on the mark. Thank heavens cap and trade did not pass our Congress and President Obama showed up later in Copenhagen with nothing to offer. No matter how you spin it, not only was a waste of time, but worse, missed an opportunity for Planet Earth and Humanity.*

On 1July09 I tackled the problem of aging. Here I was just touting eternal life, and now I'm saying we are getting too old. Both are related challenges.

Are We Getting Too Old?[237]

BACKGROUND

I am currently featuring Eternal Life in my daily report,[238] which is Chapter 2 **of Simple Solutions for Humanity**. Globally, almost 60 million die each year. This total could well drop by a huge fraction if the aging gene is found and checked. Developed countries are already getting old. Eternal Life could become a real problem.

As an early step, look for a full court press in the U.S. Congress on health. I expect some sensible action here, especially with the arrival of Al Franken. No, not because he is the medical savior, but because he is now the 60th Democrat in the U.S. Senate.

However, wait till they need to later re-look at retirement, as the 65 age requirement was enacted in 1935 when the life expectancy was 62.[239] Yes, in 1983 an adjustment was made to increase this age to 67 by 2027, but by then the life expectancy will be 82. There is a significant disconnect here with reality, but this condition is also encouraged by a private sector desiring better educated and cheaper youth over less productive and expensive elders.

POPULATION

While the world population today is 6.8 billion,[240] you might find interest in how we came to be. It is estimated that there have been about 100 billion of us since a little more than 150,000 years ago when Homo sapiens appeared. Halfway into our existence, about 74,000 years ago, we almost went extinct when Mt. Toba, another Sumatran volcano, erupted, keeping the world "dark" for six years. Our population dropped to, perhaps 1,000. As recently as 10,000 BC, there might still have only been a thousand of us, or maybe as much as 10,000. There are no census reports. Estimates vary, but there were probably several hundred thousand humans around the time of Jesus Christ in the Year Zero.

The first billion was reached in 1804, second in 1927, fourth in 1974 and the next doubling to 8 billion is expected to occur in 2025. Nine billion is now looming for around 2050 or later.

---

237 http://www.huffingtonpost.com/patrick-takahashi/are-we-getting-too-old_b_223579.html

238 http://planetearthandhumanity.blogspot.com/2009/05/well-today-i-start-excerpting-chapter-2.html

239 http://www.infoplease.com/ipa/A0005148.html

240 http://www.census.gov/main/www/popclock.html

We might never reach 10 billion for reasons to be presented. However, considering resource availability, the ideal world population should certainly be far less than the current 6.8 billion. Some doomsdayers say it is already too late.

Our population reached two peaks: 2.2% annual growth in 1963 and 163 million births about a decade ago, now already down to 137 million/year. This growth rate, fortunately, has been halved today.

China has the highest population with 1.3 billion, India is #2 at 1.1 billion (but will overtake China by 2030), Europe is #3 with 0.5 billion and the USA #4 with 0.3 billion. In 1900, Africa had 8% of the people and Europe had 24%. In 2150, the prediction is that Africa will have 24% and Europe 5%.

ARE WE GETTING TOO OLD?

The answer is yes for developed countries, and here is what is happening. Much of Europe is already experiencing negative population growth. Russia, at 140 million today, will drop to 110 million in 2050. Japan, now at 127 million, will sink below 100 million in 2050. Interestingly enough, the U.S. is still growing, 40% of this by immigration. Hmm, that border issue? Maybe we do need more youth. Our superpower status should be completely unchallenged by mid century because even China, because of their current birth control policy, will get old before becoming rich.[241]

Thus, another huge economic crisis is looming because, for developed countries, in the 80's, 5 workers supported a retired person. This has dropped to 4:1 and will crash to 2.2:1 in 2050. The current **Economist** has a special report on aging populations, and much of the following is extracted from that feature.

Let's look at China, for if they get serious about one child per family, this individual will have two parents and four grandparents to support, or 1:6. Note that this ratio has reversed. Of course, China does not have that strict a policy and some of the grandparents will not be around. In any case, the ratio could well be less than 1:1. The two countries we most toss around as current and future enemies, China and Russia, will thus have huge internal problems over the next few decades. The time might now already be at hand to shift our lopsided national security expenses toward more humanitarian pursuits in anticipation of their decline. Iran and North Korea will not stimulate a nuclear winter.

Yes, we still are wrestling with our economy, and coming down the pike are Peak Oil and Global Warming. If the American Clean Energy and Security Act[242] is a sign of how effectively we are considering these issues, I shudder to think what will happen to aging, for more than

---

241 http://www.scribd.com/doc/23446960/Will-China-Grow-Old-Before-Getting-Rich

242 http://www.huffingtonpost.com/patrick-takahashi/the-carbon-dioxide-credit_b_221933.html

half of our population will be older than 50 by 2025. In particular, watch out for AARP (formerly known as the American Associated of Retired Persons).

The whole point of the above is that decisions on retirement age must be made, if not now, very soon, or the politics of the situation will make necessary decisions impossible. Oh, Eternal Life? Never mind.

Comments (10): The discussion was varied with some dooms day threads. I guess people agree in general that our planet already has too many for sustainability. I guess immortality, then, will only add to the problems. Many challenges lie ahead.

This was by the far my most difficult and agonizing posting, ever. My wife had just passed away late the morning of 19July09, and I returned home to my computer and swiftly created "Gratitude...Not Grief," which the **Huffington Post** *published the next morning, 20July09.*

Gratitude...Not Grief[243]

My best friend, and wife, of almost 47 years, Pearl, just passed away, and it was both the saddest day of my life, and, surprisingly enough, still, joyfully blissful. We had discussed at some length the matter of death, and we both agreed that there should be no mourning period, but instead, a rebirth for the survivor. My blog of July 16[244] provided "Some Thoughts on Coping with Uncertainty and Despair." Now, the uncertainty and the despair are gone, and Pearl is at peace. Thornton Wilder, perhaps, said it best:

"The highest tribute to the dead is not grief but gratitude."

During the past month, Pearl suffered from BOOP,[245] a clot in the artery to her lungs and MRSA.[246] (For those new to blogs and such, all you need to do is click on those colored words, and you will be transported to another site providing the details.) While she was uncomfortable throughout the ordeal, she was sedated, so should not have suffered much pain. However, every day, her support structure, people like me, couldn't help but associate her treatment with what was happening around us, and most of us weakened with each passing day.

She was on nearly constant intravenous feed of Heparin and Propofol for much of this period. There remains, of course, the Heparin[247] (blood thinner) controversy, where 82 died a couple of years ago from what is still not confirmed, but is suspected to be contaminated samples from China containing chondroitin sulfate. Doubly worse for Pearl is that sometimes this compound comes from crustacean shells, and she is allergic to crab, lobster, etc. Particularly unnerving was that Michael Jackson might have died from Propofol[248] (a sedative).

As I mentioned in that previous blog posting, I'm now covering the medical aspects of my book 2, **SIMPLE SOLUTIONS for Humanity**, and just a few days ago I indicated that there are more deaths (almost always caught in a hospital) from MRSA than AIDS in the U.S.

---

243 http://www.huffingtonpost.com/patrick-takahashi/gratitudenot-grief_b_241390.html

244 http://planetearthandhumanity.blogspot.com/2009/07/some-thoughts-on-coping-with.html

245 http://www.epler.com/BoopWhat'sBoopDiseaseInformation.htm

246 http://www.mayoclinic.com/health/mrsa/DS00735

247 http://www.heparin-legal.com/news/2009/05/22/fda-clears-baxter-in-deaths-following-heparin-injections/

248 http://articles.chicagotribune.com/2009-07-21/news/0907200775_1_michael-jackson-death-diprivan-propofol

Finally, health care reform has dominated the discussion in D.C., further linking her case with the real world.

President Obama remarked that much of medical costs comes at the end of life. I can imagine if this happened to one of those 47 million or so without a medical plan. It would have been financially ruinous. Of course, you want the doctors to run every test they can to save her life, and I can only thank them for doing everything possible for her. These expenses for Pearl must have reached a quarter million dollars in a few short weeks. Is this good or bad? Well, that's part of the debate.

Anyway, life begins again for me, and do feel refreshed. Please join me in thanking Pearl for making our lives so much better. She would want me to end on a positive note, so I again quote Thornton Wilder:

"The highest tribute to the dead is not grief but gratitude."

Comments (10): The responses were touching and appreciated. While death is always difficult, there is a kind of rebirth that opens new vistas.

*I wish **HuffPo** is able to tell me how many clicked on my postings. I have no idea who I am reaching. At least my daily blog gives the exact number and home country. Also, when I submit any article, I have no sense of what the reaction might be. The matter of a substitutet for jet fuel or replacement of the current jetliner is a life or death issue for a location like Hawaii, so dependent on tourism. We will enter a prolonged state of local depression when the next quantum leap occurs in the price oil and stays there. At this point, we know what will happen, but have not taken any real precautionary steps. Thus, my 28August09 article on the future of aviation.*

The Future of Sustainable Aviation[249]

Hawaii just celebrated our 50th birthday as a State in the Union. Just about two years after attaining statehood, Barack Obama was born in Honolulu. Today, he is President of the United States, and, thirty-seven years his senior, Daniel Inouye, is Chairman of the Senate Appropriations Committee. Arguably, Hawaii, then, has the two most powerful elected officials in the Nation.

Perhaps because we are the most isolated population center on Earth, we have the highest life expectancy of any state. Our weather is ideal, we are a model for racial harmony and are, truly, nice.

Like anywhere else, though, we are suffering from the global economic collapse, with politics and personality prevailing over working together for a common cause. Our unique problem is that our revenue base is almost totally dependent on tourism. We have tried, but have failed at diversifying our economy.

Most of my **HuffPosts** have been on energy and the environment, on Peak Oil and Global Warming. There is every reason to believe that at some point in the future, say, 10 to 20 years, and, perhaps, much sooner, the price of crude oil will zoom past $150/barrel. Jet fuel will become so expensive, that tourists will stop coming to our state. Hawaii will become the first state to enter into a prolonged depression.

It's not that we have been totally asleep. In the mid-70's, studies were performed by the Hawaii Natural Energy Institute, advocating the development of next-generation aviation systems, including the hydrogen jetliner. I thus went to work for U.S. Senator Spark Matsunaga in 1979 and drafted the first hydrogen legislation introduced in the Senate. Funds for hydrogen were zero then, but last year exceeded the solar budget. The National Aerospace Plane, which was supposed to be hydrogen powered, was initiated by the Department of Defense, but today remains a black (secret) program. Why hydrogen? Because it provides the highest energy per

---

249 http://www.huffingtonpost.com/patrick-takahashi/the-future-of-sustainable_b_270969.html

unit weight, and can be produced from renewable sources, such as wind power, geothermal energy and ocean thermal energy conversion.

In parallel, we have long been researching the potential of biofuels, including jet fuel, from algae. Unfortunately, funding for this field has been limited and, while a few start-up companies are promising to produce $1/gallon fuels from microorganisms, specialists in this field tell me that we are a decade away, at best, and maybe never, of being competitive with fossil fuel options.

The Defense Advanced Research Projects Agency will spend $100 million over the next few years to evaluate this field, but the reality is that we should be spending a billion dollars annually. Three-quarters of the energy used by our military goes to jet fuel.[250]

While there are several fanciful air systems bandied about in various publications, one in particular, the Hawaiian Hydrogen Clipper, a hydrogen-powered dirigible potentially capable of flying at 350 MPH (none of the other blimps go anywhere close to this speed), proposed by Rinaldo Brutoco, President of the World Business Academy, I think shows the most promise. In particular, he sees Hawaii as the ideal lead for this effort. (This concept is mentioned in my *Huffington Post* article of December 18.[251])

There is general consensus that the following represents where the Nation and World stand regarding clean energy:

1. There has been good progress on electricity from renewables.

2. Sustainable ground transport options are in advanced stages of development.

3. Sensible next generation aviation systems have been largely ignored.

As Hawaii is the site in greatest jeopardy, with high interest from the military, it should be justifiable for Senator Inouye and President Obama to provide $1 billion/year to develop the National Hydrogen Clipper through the Department of Defense. This should become a program for international collaboration, for the whole world will soon, also, be in trouble if nothing is done about the future of aviation.

Comments (43): Incredibly, there were 43 responses. However, most might have been unrelated to the issue, and others beyond the pale. Maybe, though, enough people read it to help make a long-term difference.

---

250 http://www.eenews.net/public/climatewire/2009/07/27/1

251 http://www.huffingtonpost.com/patrick-takahashi/a-solution-for-barack_b_151822.html

A campus colleague, Jay Hanson, sent me various e-mails about his America 2.0, so I expanded that concept to World 2.0 on 15October 09. The term, Government 2.0,[252] or e-Government, seems to be gaining interest as a means of redesigning our national government. The following piece again harps on the re-prioritizing our budget towards civilian applications.

World 2.0[253]

The United States is the greatest country ever, and there will be no threat to our supremacy for a century and more. The Cold War brought our civilization to the brink. But that was almost two decades ago. There are no more threats. Afghanistan, Iraq and North Korea are mere mild irritants.

What about that feared China? Their one child policy means that the country will get old before it gets rich, as vividly portrayed by **The Economist** in their 25June09 issue. Why? In 2000, ten people supported one retired person, or a ratio of 10:1. In 2050, China's could drop to 1.6:1. So let's say they change this population policy. Japan in 2050, with no such restriction, is nevertheless expected to drop to a ratio of 1.7:1.

Could Russia re-challenge? Probably not, as their maximum population, which was once close to 150 million, could well drop almost to 100 million by 2050, and will then have fewer people than Vietnam. The USA is projected to reach 400 million by mid-century.

All that said, we are not perfect. In fact, we are incredibly imperfect. I quote the final two paragraphs of my 19February09 article in the **Huffington Post**:[254]

To summarize, the majority of Americans believe in both creationism and an afterlife, the potential of some sort of religious doom, and think they are not causing global warming. So the title of this article should have been: "Creationism, Doomsday and the Afterlife," to more closely reflect life in the USA. You now should have a better understanding about why we are in deep... (feel free to add your own odious term). So what has this got to do with the economy? Go back to the beginning and try again, or revert to my earlier HuffPost introduction to this subject.

Did we become the greatest country ever because of our beliefs? Certainly not entirely, which gives me hope that the best is yet to come.

More recently, I have agonized about the inability of our Congress and President Barack Obama to take any true progressive action. This leads me to another quote:

---

252 http://www.manhattan-institute.org/government2.0/

253 http://www.huffingtonpost.com/patrick-takahashi/world-20_b_321886.html

254 http://www.huffingtonpost.com/patrick-takahashi/evolution-global-warming_b_168827.html

Thomas Jefferson, along with James Madison, worked assiduously to have an 11th Amendment included into our nation's original Bill of Rights. This proposed Amendment would have prohibited 'monopolies in commerce.' The amendment would have made it illegal for corporations to own other corporations, or to give money to politicians, or to otherwise try to influence elections.

If only Jefferson and Madison had prevailed. This quote was extracted from a paper offered by Jay Hanson. As I don't seem to have a solution, and he thinks he has, I can only urge you to read his concept entitled, America 2.0.[255]

What we really need is a World 2.0, for we do have problems -- huge, global ones. That grand recession was just an appetizer. Peak Oil and Global Warming are like twin asteroids already in close striking distance. Chapter 1 of my **Simple Solutions for Humanity** provides the beginning outline of such a concept. The problem, of course, is how to strike sufficient fear to compel meaningful decision-making. If there is no Cold War and no real world threat to our freedom, why don't we just shift all our expenditures from war to these looming cataclysms? Ah, that is my original article in the Huffington **Post** entitled, "Well, Barack, We have a Problem," posted during the heat of the Democratic Presidential Campaign.

But even if Obama and the well-meaning Democrats rode into power, the fact of the matter is that millions did not perish this summer from excessive heat and there is no discernible sea level rise. It is thus an easy prediction that our combined governments will only wave hello at Mother Nature from Copenhagen in December. That is the flaw that World 2.0 needs to overcome: society can presently only react when it is too late.

Comments (2): Another ho-hum response. People are just tired of global warming.

---

255 http://www.warsocialism.com/america.doc

*In November and December, the media widely reported on biofuels from marine algae. I thus reviewed the field on 9November10 in anticipation of these gatherings. I actually wrote a Part 2, which I submitted to the **Huffington Post**, but retracted because the tone of my paper bothered me...it was too negative. I did not want to offend researchers and officials making an honest attempt at developing the field. Sufficient time has passed, so we will see this version after Part 1.*

Biofuels from Microalgae (Part 1)[256]

The United Nations Environment Programme[257] recently published its first Biofuels Assessment Report[258] ever. Much of this assessment dealt with conventional biomass, and mostly, the report did a fine job saying some bio systems are good, some are not so, and much depends on how you do it. Global warming remediation and economics were dominant parameters, although water, state of the technology and other factors were considered.

Let me focus on what many think might be the most promising ultimate bio option. I've been surveying colleagues for several years now on biofuels from algae, and the speculations on potential cost are all over the map. But the potential is exciting, for it is said that you can grow several times (factor of two to ten, you pick a number) more biomass from an aqueous environment than on land. Mind you, this point remains debatable.

Part of the reason given is that terrestrial plants need to pass nutrients only through thin roots, defying gravity, while aquatic micro and macro species can use the total surface area. Plus, genetic engineering can more readily be applied for a micro system, which has an effective doubling time of hours, not weeks, months or years.

For this analysis, which comes in two parts, I will focus on microalgae grown in saline water on land. A follow-up article will review prospects for macroalgae (such as kelp), the form pioneered in the open ocean by Howard Wilcox as early as 1968, and now, mostly being investigated today by the Japanese. This early work mostly led to methane by fermentation and as feed for animals. Recent interest adds ethanol to the product mix. A fourth posting will blue-sky the prospects for actually attempting to utilize the effluent from ocean thermal energy conversion (OTEC) plantships to manage algal farms at sea.

My speculation is that terrestrial microalgal and marine macroalgal biofuels/feed systems are a decade away from commercialization, and only if the price of oil by then exceeds \$125/ barrel. The combination of our sun, the ocean, microalgae, OTEC, and genetic engineering for sustainable marine biofuels (hydrogen, alcohols, biodiesel, etc.) is probably a generation

256 http://www.huffingtonpost.com/patrick-takahashi/biofuels-from-microalgae_b_347093.html

257 http://www.unep.org/

258 http://www.unep.fr/scp/rpanel/Biofuels.htm

or two away. This would be an element of the Blue Revolution.[259] Thus, ultimately, there will be four postings.

I begun to be involved with growing algae in raceways a third of a century ago, and from then until now, have observed that federal funding was spotty and mostly non-existent. There was never a truly orchestrated national program and sporadic attempts at organization were thwarted by the fickle price of oil. There remain today too many unknowns and uncertainties, for the due diligence and science have not yet been performed. The fundamental engineering was never initiated, and remains a knowledge gap, for this work should proceed in parallel to someday mesh with the science. The National Science Foundation[260] for the past few decades has tended to avoid funding energy projects, mostly a jurisdictional attitude in favor of the Department of Energy, but is finally beginning to recognize this deficiency and has initiated steps to take a more active role.

So let's get to the heart of the matter regarding terrestrial microalgal biofuels production: the eventual cost of production. In general, the price of crude oil is a good an indicator as any of what biofuels from algae facilities must meet to be competitive. Let us look at the numbers. Take crude oil at $80/barrel, or $1.90/gallon. A typical USA average these days is $2.73/gallon for regular gas. The ratio is 1.43, that is, gasoline costs 43% more than crude oil. (Prices change, and on November 27, this ratio was 1.48.) This ratio was 1.64 in 2008, 1.85 in 2007 and 1.92 in 2006. The differential accounts for profits, taxes, marketing, etc., and will drop as the price of crude oil rises, unless, of course, there are added taxes.

One way of looking at this is if the best industry can do is produce biofuels for $5/gallon, then oil needs to go up to $220/barrel. If the production cost can be reduced to $3/gallon, then, oil would only need to rise to $126/barrel. My gut feeling is that $3/gallon will only be attained with considerable R&D over a period of 10 years or more, and maybe never.

However, there are various mechanisms to foster the earlier coming of biofuels from algae. One is to link the project to pollution control and a range of bio-co-products, for algae has a lot of protein, something absent in terrestrial biomass. The added value factor can make that crucial difference, and this especially becomes obvious with the Blue Revolution. A second, more ethereal, potential introduces the matter of life cycle costing, for if the financing can proceed with the confidence that oil will rise beyond $150/barrel, with the added attraction of government incentives, these operations might well attain prominence relatively soon, even if oil might only be in the range of $100/barrel.

Comments(25): *Discussion was heated, but tended to divert away from biofuels from algae to solar PV. There is a lot of hype about renewable energy in general, and algae fuels in particular.*

---

259 http://www.huffingtonpost.com/patrick-takahashi/the-dawn-of-the-blue-revo_b_145889.html

260 http://www.nsf.gov/news/news_summ.jsp?cntn_id=112243

Well, this is that article I decided not to publish in 2009. I offer it here because the field is coming to its senses and my input is not so sensitive anymore.

Biofuels from Microalgae (Part 2)

Knowledgeable colleagues tell me that microalgal biofuels today would cost about $50/gallon to produce, while a current Department of Defense estimate[261] shows a minimum figure of $20/gallon (mind you, this is $840/gallon). Let us, then, look at a few possible biofuels from algae speculative future cost of production per gallon estimations:

$ 1 .... A few entrepreneurs[262] (of significant dubiousness)
$ 2 .... Department of Energy (very unofficial, but murmured)
$ 3 .... Defense Advanced Projects Agency[263] (desired)
$ 4 .... Noted scientific authority (someday if all goes right)
$ 5 .... DARPA definition of affordable[264]
$10+ .... Noted industrial authority

My noted scientific authority, and he is the best in the field, said this is like comparing apples and asteroids, but gave $4/gallon. For now, I'll keep him anonymous. He is right, of course, for what do those above figures mean? Someday with major breakthroughs in genetic engineering? In addition, he provided a dozen more qualifiers. Well, for one, almost surely, these guesses represent the cost after a decade of development, not today. Even then, the Department of Energy projection must be more wishful than anything else. At least, though, that department is now treating this field with some urgency and has applied $50 million of stimulus funds toward this cause. DARPA, more so, is reported to have set aside $100 million for this adventure. But the Department of Defense is the single largest consumer of energy in the country and half is for jet fuel,[265] so they better be concerned. Ah, the private sector: Exxon Mobil[266] is said to have dedicated $600 million, in partnership with the genome table co-champion, Craig Venter. Regarding the $10+/gallon quote from the noted industrial authority, he did, in fact, say I could use his name, but won't.

So to summarize, don't believe $1/gallon biofuel from algae, hope for $3/gallon someday, but for the next few years, don't be surprised if biofuels from algae only become competitive when oil reaches $200/barrel ($4.76/gallon). As mentioned in Part 1, linkage with pollution control or the co-product of animal feed can provide an added value factor, while generous

---

261 http://www.chiefengineer.org/content/content_display.cfm/seqnumber_content/3236.htm

262

263 http://www.motherjones.com/politics/2009/09/algae-energy-orgy

264 http://e85.whipnet.net/alt.fuel/algae-oil.html

265 http://www.energybulletin.net/node/29925

266 http://www.motherjones.com/politics/2009/09/algae-energy-orgy

tax incentives, as for example, presently available for ethanol, plus incorporation of life cycle costing and externalities, would work. For their $600 million investment, you can bet that Exxon Mobile is covering its future by setting the stage for their success through traditional Congressional and White House discussions.

The situation seems more difficult for jet fuel,[267] as the current price is about $2/gallon, about the cost of crude oil itself. It is more expensive to refine jet fuel than gasoline, yet, over time, the selling price of jet fuel has been cheaper than gasoline. The ratio for jet fuel[268] has over the past few years been in the range of 1.25. How can this be? Well, bulk purchases, advanced commitments and lower taxes. In any case, a microalgal jet fuel producer, thus, will actually be faced with the same production cost to match crude oil, as one selling biogasoline, for the price to the consumer is, for the investor, almost irrelevant.

The State of Hawaii absolutely depends on DARPA succeeding beyond all expectations, for at those predicted astronomical future oil prices, which could come at any time, and certainly in five to ten years, airline fares will go sky high, tourists will stop coming to our state and we will enter into a prolonged great depression. Unless, of course, the Hawaiian Hydrogen Clipper,[269] by some miracle, suddenly gains an Apollo-like following, with mushrooming wind farms, geothermal fields and OTEC plantships providing cost-effective hydrogen. Yes, dreaming... but not much more so than the early commercial reality of bio jet fuel from algae.

---

267 ="http://www.iata.org/whatwedo/economics/fuel_monitor/index.htm

268 http://www.airlines.org/economics/energy/Annual+Crude+Oil+and+Jet+Fuel+Prices.htm

269 ="http://www.lifenotnews.com/mikeflynn/index.php?option=com_content&task=view&id=83&Itemid=1

*The term, national debt versus gross national (or domestic) product, can sometimes be confusing. This has mostly to do with what is considered to be debt. The CIA, for example, only uses debt held by the public, thus, their debt/GDP ratio is always very low. In any case, at the end of World War II, the ratio used by most sources (gross debt) rose to above 120%, then slowly dropped until the second energy crisis of 1979. The prognosis is for our debt matching GDP over the next few years, with both being in the range of $14-17 trillion.[270] My **HuffPo** of 23November09 stepped into this tricky field.*

How Serious Is Our National Debt?[271]

In answer…maybe not as serious as you might think. Let me tell you why, looking at three sources, beginning with the October 24, 2009 issue of **The Economist**.[272] The issue shows a projected American government debt as a percentage of our Gross Domestic Product (GDP) of about 100% in 2010. However, Japan will go to 230%!

Interestingly enough, the **CIA Factbook**[273] has a table of essentially the same international comparison, but for 2008, with Zimbabwe at the top with a debt percentage of 266%. Japan is listed #2 at 172%. However, I couldn't find the USA until I reached #61 at 38%.

So what's going on? Easy, the CIA uses only public debt, not gross debt. Let's do a simple calculation. Our national (gross) debt is just over $12 trillion. Our GDP is $14.4 trillion, which results in a figure of 83%. Thus, in 2008 our national debt of about $10 trillion resulted in a % of GDP ratio of 70%, with a 2010 estimate of $16.6 trillion debt and 98.1%.

Thus, yes, the American debt as a percentage of GDP is at around 100%, which is expected to edge up to 101% in 2011, then begin to drop. The expected continued low interest rates can only help a debtor, our government.

Expect, though, a policy uptick, for the health care measure will initially add more debt (just plain old common sense as 35 million or so more people will now need to be covered), but within the decade, when the public option finally kicks in to truly compete, the unacceptable growth rate of our national medical bill should be checked. If you're rich, you'll be hit twice: you'll subsidize much of this, and, so that you can cut in line for service, you will buy supplemental insurance.

Oh, by the way, also from **Wikipedia**:[274]

---

270 http://en.wikipedia.org/wiki/United_States_public_debt

271 http://www.huffingtonpost.com/patrick-takahashi/how-serious-is-our-nation_b_366811.html

272 http://www.economist.com/node/14699754

273 https://www.cia.gov/library/publications/the-world-factbook/geos/us.html

274 http://wapedia.mobi/en/United_States_public_debt

The debt limit was most recently raised to $12.104 trillion by the American Recovery and Reinvestment Act of 2009 (H.R.1), which was signed into law on February 17, 2009 (P.L. 111-5). [11]

Is there a check and balance system in place or can our national debt keep going up forever? The answer is yes, for Congress needs to approve it, and yes, again, because it always does when asked by the President.

To go on, a look at a historical graph (from **zFacts**)[275] shows that the all-time national high of 120% was attained at the end of the Second World War, but the current exponential slope looks damning. However, 83% or 100% is still nowhere close to Japan, which appears to be surviving at double our rate. I should mention without going into details that the Gross Domestic Product is about 10% lower than the Gross National Product, and the reason why we don't have exact agreement among sources is because of this discrepancy and the year being cited. Then there is the CIA with only public debt.

This same graph shows that the Reagan-Bush (Senior) reign showed a doubling of our national debt/GDP percentage, while the Bush (Younger) years initiated the jump when Congress passed the initial bailout package in December of 2008 before Obama came into office. While we're at this, you should know that President Reagan, when he came into office in 1982 faced exactly the same predicament as Obama, for the second energy crisis in 1979 had discombobulated the economy. In 2009 dollars, Reagan got a $1.8 trillion recovery package, double that of Obama.

Finally, what about China? That same **Economist** article reports that China does own 24% of our foreign debt, but that Japan, a country with a 200% or so debt/GDP percentage, is at 20%. Actually, China recently dropped to 23% and Japan rose to 21%, and in 2007 this was not as bad as you might think, as foreigners only then accounted for about 25% of our national debt. 75% was owned by us. Thus, China's hold on our economy was actually less than 6%. <u>But foreign ownership of our national debt has doubled, so China's grip is now at 11%.</u>

But let them them bolt and invest in Zimbabwe (remember, their debt/GDP percentage is 266% -- and China has a platinum problem with this country today) instead. Yup, it is appearing that China is contemplating moving money from the U.S. into African, South American and Indonesian resources. It's a risk, but, think about it, would you rather trust the U.S. economy or gain sure access to world resources, which will only jump in prices over the next decade?

So, be mildly concerned about our escalating national debt, but there is no need to anguish. Consider Japan. Also, mostly ignore those editorials that regularly pop up throwing fear at you about China pulling out their money, causing an American depression. They probably will reduce their trust in our economy, but could run into greater difficulty dealing with many of those developing countries that now and then tend to nationalize their industries.

---

275 http://zfacts.com/p/318.html

Comments (23): There was a lot of discussion, all very educational. I learned a lot. Sometimes I wonder why I even comment on the economy for I don't think I have even taken Econ 101. Yes, maybe I should return to energy and the environment.

Thanksgiving was soon to come, so on 23November10 I thought I'd share my experience on roasting a turkey, something I had never done before. The final product was outstanding, but eight months later I've only managed to eat two of the eight ziplock bags of turkey and stuffing. I wonder how long they can be kept in the freezer?

How To Roast A Turkey[276]

*[In July, the day after my wife passed away, The **Huffington Post** published my article on "Gratitude, Not Grief."[277] With all the trauma now fading memories, I have entered a new and rather exciting phase of my life.]*

For some of you looking for a reasonably safe adventure, you might want to consider roasting a turkey for the holiday season. I would like to share with you my first attempt.

I noticed that Safeway was charging only $3.99 for any turkey 16 pounds or less... not per pound, but per whole fowl. (Later I learned that various supermarkets also had similar sales, and, for all I know, this happens every year at this time. I would imagine that a state like Minnesota, which is noted to produce the most number of turkeys--the kind you eat--must give them away for free. Oh, another nice piece of trivia is that the US annually raises just about as many of these birds as our total population.)

For that price, I would have been satisfied with a pigeon-sized bird, but the smallest one I could find was 11 pounds. Then, I couldn't check out because the fine print said I had to have a bill of at least $20. But that was no problem, since I bought a few more necessary items. Unfortunately, at home, I couldn't fit it into the refrigerator, so I placed it into one of those insulated bags for defrosting.

The next day I learned from my golf group that it would be smart to first soak the bird in a saline bath to kill the salmonella and such, and add some taste. I did not have enough salt, so again I went to the market and bought a whole standard model cylindrical container of salt for all of $1.07. Amazing, considering the one pound 10 ounce weight and shipping cost. I then thought about cranberry sauce and sweet potato, but, no, it was not really that day yet, and I would see too much of it on November 26, so I went home with only the salt.

However, circumstances prevented my actually placing the bird in the oven at that moment, so I added ice as necessary. Already it was much more work than I wanted, especially as I don't particularly like turkey.

---

276 http://www.huffingtonpost.com/patrick-takahashi/how-to-roast-a-turkey_b_366778.html

277 http://www.huffingtonpost.com/patrick-takahashi/gratitudenot-grief_b_241390.html

Rather than going to the Internet, I noticed a large brown Treasury of Great Recipes in the kitchen, and found "Roast Turkey Wayside Inn." I hate cloves, parsley, thyme, neck, heart, liver and giblet, so I purposely left them out. Oh yeah, you need to remove those organs inside of the turkey. There was also a plastic contraption which served no particular function to me, and maybe could melt in the oven, so with great difficulty, I removed it. Maybe a reader will comment on this matter.

I'm also not a great fan of bread stuffing, so I created my own: cooked rice, can of corn, water chestnuts, and chopped macadamia nuts/mushrooms/onion. I found some bacon, and with some irony noted that it was made of turkey. So that's what Pearl was feeding me. Anyway, I crisped the bacon and worked it in with a raw egg, plus some salt and pepper. The whole concoction perfectly fit into the turkey and I tied the legs to keep everything in place, barely. Into a large pan with aluminum foil lining, I added two cups of water and a quarter pound of butter.

The main parameters of importance I sought from the book were what temperature (325 degrees F) and for how long. To my chagrin, stated was: ten to twenty minutes per pound. Thus, using a calculator I determined that the oven should be on for anywhere from less than two hours to nearly four hours. I arbitrarily selected three hours. The main thing was to gain the right shade of brown, which was a slam dunk, as the other options are white and black.

The directions called for basting every half an hour. Basting? What's that, and how? Well, that's somehow getting the liquid part in the pan spread over the bird to keep it moist and, ultimately, tastier. I think I needed that bulb and tube thing, but a large spoon sufficed. Be careful, as this can be a dangerous process when the oven is hot.

Well, three hours later, perfecto. After a cooling period, I didn't bother with careful carving because no one was watching and I also never learned. Instead, I cut delectable portions and placed them on a plate. The recipe also called for gravy, but why bother with having to wash another pan and add flour. It would be sacrificed anyway if diet was a factor. I placed a bit of pan liquids over some mashed potato I found in the freezer from a previous experiment, had the outstanding stuffing a la Takahashi and fixed a lettuce and tomato salad with blue cheese dressing. Complemented by a glass of cabernet sauvignon, no, make that two, I had one of my best meals, ever.

You know, maybe now, I might begin to appreciate turkey, which is a good thing because the leftovers filled eight quart size Ziploc bags, and should last me the lifetime of the freezer. For only $70 - $150 I could have bought a whole prepared turkey with all the trimmings from assorted suppliers, but this was for the sheer experience, plus it was incredibly economical. Heck, I even had more than 50 cents of salt left for future use.

My next adventure could well be a goose for Christmas...or, maybe Peking Duck.

Comments (4): Not much in way of comments, but the experiment was enjoyable.

*I knew nothing about thorium until reading **Wired** on the subject, but was so intrigued that on 5January10 **HuffPo** published this paper. Interestingly enough, the hero of the article, Kirk Sorensen, who operates a thorium energy blog,[278] read my **HuffPo** on thorium and communicated. Turns out he also is a fan of ocean thermal energy conversion. So we'll be interacting on both.*

There is Something About Thorium[279]

Just about a year ago, I wrote an entry on the **HuffPost** entitled "Renewable Electricity is Our Only Viable Option"[280]. In it, I panned both nuclear and coal fuels for a variety of reasons.

I need to make a mid-course correction about nuclear power.

There are two types of nuclear energy:

- *Fission*, like that embodied by the Atomic Bomb, where a special isotope of uranium and/or plutonium is used as the fuel, and

- *Fusion*, like the Hydrogen Bomb, where isotopes of hydrogen are combined to produce energy.

The light and heat from our sun and all stars result from the fusion of hydrogen. Thus, of course, solar energy is also nuclear energy.

The first type of nuclear energy, fission, results in radioactive wastes that have the potential to remain dangerous for many hundreds of thousands of years. The second, fusion, similarly has to be stored, but only for several tens of years.

Uranium/plutonium fission materials have the potential to be exploited by terrorists for dirty bombs. Further, depending on who you ask, there is such a thing as Peak Uranium. France, for example, now only imports its nuclear fuel. The price of uranium is also fickle, which for decades dawdled at $20/pound, only to zoom up close to $140/pound in 2007 before settling more recently at $45/pound.

Remembering Hiroshima/Nagasaki and Chernobyl/Three Mile Island, in context of nuclear terrorism combined with the sticky problem of where to store nuclear waste, I have long been an opponent of embracing fission. I still take pride in helping kill the Clinch River Breeder Reactor when I once worked as a staffer for the U.S. Senate.

---

278 http://www.takeonit.com/expert/209.aspx

279 http://www.huffingtonpost.com/patrick-takahashi/there-is-something-about_b_410825.html

280 http://www.huffingtonpost.com/patrick-takahashi/renewable-electricity-is_b_162435.html

However, a special form of "clean" fission seems to be making a comeback. ***Cosmos*** 2006[281] and the latest issue of ***Wired*** both cover thorium as an option for nuclear electricity.[282] Named after the Norse God Thor, the element shows exciting potential as fuel for a "greener" next-generation fission reactor.

Earlier this year I was disappointed to learn that Abu Dhabi, the creator of the sustainable Masdar City,[283] with all that natural sun, was jumping into nuclear power.[284] But I now better understand that Masdar will be carbon-low, and nuclear is certainly one pathway. Further, I was pleased to read that Thorium Power[285] gained a small contract to advise the country.

You can click on the above links to gain a more in-depth knowledge on the subject, but let me give you my top ten reasons (not in any particular order, and a few are counterintuitive, if not shocking) for advocating thorium, at least to promote a thorough public discussion:

1   There is up to six times more accessible thorium than uranium,[286] with the U.S. being second to Australia. Others say India has one-third the resource and there is thorium to supply all our needs for a thousand years, but that's the ever changing nature of a newly developing concept.

2   The very first commercial nuclear powerplant, Shippingport, was powered by thorium during the final five years of operation, ending in 1982, so we know the concept works. However, America had earlier because of the Cold War turned to uranium/plutonium because our "war" advisors wanted this fuel for nuclear weapons. Just this decision pushed the world to the brink of nuclear winter and now provides the ingredients for a dirty bomb.

3   On an annual basis, for a typical 1000 MW uranium powerplant, you start with 250 tons of uranium ore. A 1000 MW thorium system uses one ton of thorium, and the typical ash produced in a 1000 MW coal facility results in 13 tons of thorium.[287]

4   The fuel cost for a conventional nuclear powerplant is $50-$60 million, while the equivalent thorium reactor will only use $10,000 of thorium.

5   Uranium/plutonium wastes need to be safely stored for hundreds of thousands of years, while thorium is not fissile, and the reactor wastes would require, perhaps, caring "only" for several hundred years.

---

281 http://www.cosmosmagazine.com/node/348/

282 http://www.wired.com/magazine/2009/12/ff_new_nukes/

283 http://www.masdar.ae/en/home/index.aspx

284 http://online.wsj.com/article/SB123862439816779973.html

285 http://gulfnews.com/business/general/thorium-power-to-assist-uae-in-nuclear-programme-1.107237

286 http://www.world-nuclear.org/info/inf62.html

287 http://newenergyandfuel.com/http:/newenergyandfuel/com/2009/01/21/the-liquid-fluoride-thorium-paradigm-part-i-thorium/

6 There is no terrorism potential for the thorium cycle. There can be no nuclear meltdown for a thorium reactor.

7 Uranium fission, thorium fission and fusion produce very little carbon dioxide (and not from the process itself, but from the materials and during construction).

8 Regarding the size of land required, a 1000 MW nuclear power site needs about a quarter million square feet, surrounded by a huge buffer zone. A thorium 1000 MW facility would only need 2500 square feet, with no buffer zone. I'm just reporting from the **Wired** article above. You might not be able to legally build a house on a lot this small.

9 Both Canada assisting China[288] and India[289] are rapidly advancing thorium fission.

10 Senators Orrin Hatch and Harry Reid have introduced legislation for the thorium cycle.

My book well covers nuclear power and I worked at the Lawrence Livermore National Laboratory on laser fusion. I don't previously remember the thorium option being even discussed. Yes, it's not perfect because you still need to mingle thorium with some rare uranium and there are assorted warts, but if global warming is accepted as real, we immediately will need viable options to replace coal, and the thorium fission reactor should as soon as possible undergo comprehensive due diligence, with step two being to replace pure uranium in a few existing nuclear power facilities (yes, the retrofitting option is another bonus) with the thorium mix.

Several environmental groups have embraced nuclear power in light of global warming. Could thorium be that magic bullet to bridge humanity over the next century or more to provide cleaner electricity?

Comments (12): The discussion was uniformly supportive. Hmm…maybe there is something to thorium. Here is one of my responses:

Can you believe it, Marcel? I'm actually promoting fission energy. I did get a few catty remarks from some of my green friends. I haven't yet, though, received any compelling arguments to sway me away from this option…yet. I even purchased a few shares of Lightbridge (which is the new name of Thorium Power). Unfortunately, it promptly dropped 5% the first day. Doesn't matter, for I'm in it for the long haul.

---

288 http://www.nuclearcounterfeit.com/?tag=thorium-fuel

289 http://www.indiadaily.com/editorial/17398.asp

*In January I began my around the world adventure, with a special stop at the Taj Mahal planned to toss Pearl's ashes. She always wanted to visit that attraction and never made it. I've initiated a crusade to go where she wanted, and leave part of her there. South Korea was my first country, and **HuffPo** on 20January10 published my report.*

The Wonder of South Korea[290]

I've embarked on a world journey, perhaps my tenth, to learn more about Planet Earth and humanity. My first stop is Seoul, South Korea. I've been here at least annually for the past quarter century, and sometimes twice or thrice a year. I would like to share some current insights and gain the views of the **HuffPost** readership through your comments.

With regard to the region's economy, it has recovered remarkably well. While we are stuck at 10% unemployment, similar to that of Europe, the unemployment rate of Singapore is 3.4%, South Korea 3.6%, China 4.3% and Japan 5.2%. Exports for Singapore, for example, jumped 26% last month from the previous December.

However, all is not necessarily well. A November article in **The Economist** about China getting old before it gets rich was reinforced by the **Korea Times**' graying report of China published today.[291] The developed countries in this region are now beginning to experience a decline in their populations.[292] The governments of South Korea and Japan have initiated steps to provide incentives for families to have more children.

The suicide rate in South Korea,[293] now at 26 per 100,000, is the highest of developed nations, even though those of the former Soviet countries are worse. The U.S. is at 11.1, while Mexico is at 2.3. I asked why? People in Korea are now more successful, but are not happy, they say. The stress begins in their educational system and worsens when they begin to work, leading to family problems. The relative reality, though, is that, with a score of 54 on the Happiness Index,[294] the country ranks in the upper middle range, compared to Costa Rica at 79 and Tanzania at 19.

Mind you, students in the Orient are at the top of their international class on achievement, part of the reason why their homelands are so successful.[295] But in Singapore, they have lost that certain humanitarian quality.

---

290 http://www.huffingtonpost.com/patrick-takahashi/the-wonder-of-south-korea_b_429468.html

291 http://www.scribd.com/doc/23446960/Will-China-Grow-Old-Before-Getting-Rich

292 http://news.mongabay.com/2009/0519-hance_populationasia.html

293 http://english.hani.co.kr/arti/english_edition/e_national/158160.html

294 http://www1.eur.nl/fsw/happiness/hap_nat/findingreports/RankReport_InequalityAdjustedHappiness.htm

295 http://www.bc.edu/bc_org/rvp/pubaf/chronicle/v5/N27/timss.html

More specific to Korea, there is a North and a South. Click on "Planet Earth and Humanity"[296] for a more detailed summary, but the North is somewhat larger than the South, with slightly less than half the population and about one-fifteenth the GDP/capita.

President Lee Myung-bak, is said to show "Makgeolli" leadership,"[297] that drink being a once popular cheaper option for Korea University (his alma mater) students than the pricier beer drunk by archrival private school Yonsei University. But more positively, his demeanor is down to earth and personal.

Kim Jong Il, the less than great leader of the North, shows symptoms of an ailing trapped mouse in critical transition. He could well take irrational action, not unlike a junkie mugger in need of a fix or those 9/11 terrorists driven by religious fanaticism. They are are all metastable, not really crazy and dangerous. Yet, this is not a particularly worrisome factor to a Southerner, for the latest threat of sacred retaliatory battle[298] a few days ago only made page three of the local newspaper.

Anyway, it is clear that South Korea does not today want to take on the financial burden of the North, unlike what West Germany seemed to have embraced with East Germany. The South will continue to tolerate regular, and mostly, empty, threats, with prudence and patience, plus as much complementary aid and involvement from other countries, too. A sensible plan, awaiting the inevitable.

The key sore point facing relations has to do with nuclear energy and proliferation. The North is ready to cave in and the South is surging, for, against all odds, they beat the world to a $20 billion contract with Abu Dhabi to build 4 large nuclear powerplants. I discussed the prospects of thorium[299] being introduced, and found high interest.

Korea will for the first time host the G20 finance ministers gathering next month focusing on the world economy and maybe even a bit about the environment. I discussed the prospects with planners about something like the Green Enertopia[300] concept, but did not get too far. The ministers will meet in Incheon, the location of the "new" international airport, rated the best of 2009,[301] with Hong Kong #2 and Singapore #3.

Japan Airlines filed for bankruptcy this week. Part of the reason is that Japan, about the physical size of California, has nearly a 100 airports, and most of these domestic routes are highly unprofitable. Landing/usage fees in Japan average 9%, while Korea is at 3%. Thus, Incheon Airport has become the gateway to Japan.

---

296 http://planetearthandhumanity.blogspot.com/2009/02/what-about-north-korea.html

297 http://www.koreatimes.co.kr/www/news/nation/2010/01/117_59386.html

298 http://gulfnews.com/n-korea-threatens-to-break-off-dialogue-with-s-korea-1.568787

299 http://www.huffingtonpost.com/patrick-takahashi/there-is-something-about_b_410825.html

300 http://planetearthandhumanity.blogspot.com/2009/10/project-green-enertopia.html

301 http://newterminal.blogspot.com/2009/06/seouls-incheon-number-one-for-passenger.html

160

So South Korea, while too cold for me in January, is doing exceedingly well. What about their future? In light of their still strong, but declining, shipbuilding capability, as they have no resource base, I suggested that they re-invent themselves by taking advantage of the unique fact that, unlike the USA and Japan, they actually have the equivalent of the Department of the Ocean, and build the platforms for the Blue Revolution.[302]

What also about leapfrogging over everyone else by designing the next generation hydrogen aircraft? Or, maybe a swift dirigible like the Hawaiian Hydrogen Clipper.[303]

About their happiness? I have no clue.

Comments (4): Not much of a response from the reading world.

---

302 http://www.huffingtonpost.com/patrick-takahashi/the-dawn-of-the-blue-revo_b_145889.html

303 http://planetearthandhumanity.blogspot.com/2009/08/on-fractional-hydrogen-heavy-ion-fusion.html

Progress in Southeast Asia[304]

This is part two of my world odyssey, the first report being on "The Wonder of Korea".[305] Today, some thoughts combining Vietnam, Cambodia and Thailand. Details with photos can be found at Planet Earth and Humanity.[306]

When France colonized Indochina, there was a general feeling that the Vietnamese were hard workers, Cambodians hard watchers and Laotians hard sleepers. Yes, this is a joke, but many times stereotyping is based on something. Certainly, those Vietnamese that relocated to the U.S. made an impression. There was a period when statewide spelling bee champions and science awards went to Nguyens, Trans and Phans. Their children are also doing well.

If your vision of these countries is limited to the war, the ***Killing Fields*** and ***The King and I***, I've got news for you. There has been change, and some of this has been remarkable.

The G-20 major economies already include India, Mexico, Turkey and Argentina. Will Thailand and Vietnam become part of the G-25? Singapore and Costa Rica have populations below 5 million. Thailand has more than 60 million and Vietnam is up to 90 million. Interestingly enough, the latter is increasing at a rate of 10 million per decade, and it is predicted that Vietnam's population will exceed that of Russia within the next generation.

To become a world power, Vietnam has to significantly improve its economics. The per capita GDP is only a bit more than $1,000, but this is 400 percent higher than only 15 years ago. The trend is promising. China Beach (picture 100 yards of beach extending for 19 miles, a distance from Waikiki to Pearl Harbor and back to Waikiki)[307] alone has a five-star mega resort, with five more being built, including a J.W. Marriott and a Hyatt, plus a casino and two golf courses.

The joker in this deck is China. Remember Pol Pot and the Killing Fields?[308] You ask why did this happen? The people of Cambodia say that this was an attempt to exterminate them to be replaced with Chinese Han, something some say still might be occurring in Western China,[309] and, perhaps, too, Tibet.[310]

---

304 http://www.huffingtonpost.com/patrick-takahashi/progress-in-southeast-asi_b_446883.html

305 http://www.huffingtonpost.com/patrick-takahashi/the-wonder-of-south-korea_b_429468.html

306 http://planetearthandhumanity.blogspot.com/2010/02/top-ten-in-southeast-asia-south-korea.html

307 http://planetearthandhumanity.blogspot.com/2010/01/most-amazing-beach-in-world.html

308 http://www.cbsnews.com/stories/2000/04/15/world/main184477.shtml

309 http://www.nytimes.com/2009/07/06/world/asia/06china.html?_r=1

310 http://www.independent.co.uk/news/world/chinese-poor-invade-tibet-1102835.html

Any good excuse, and Vietnam could well become a province of China. The Han Chinese strategy, though, is not at work here, as the ethnic Chinese population only amounts to 3 percent, and many of them actually left Vietnam after the 1980 war with China.

This is pretty heavy stuff, so let me close with some upsides. First, traveling through Vietnam and Cambodia, life is good and there is no pervading sense of being under the thumb of communists, who run the government. Granted, my lifestyle is not quite that of Henri Mahout, who in 1860 tromped through the jungles to "find" Angkor Wat. http://www.mapsofworld. com/travel-destinations/angkor-wat.html The Angkor Thom[311] area, incidentally, near a millennium ago, had the largest population of the world at 1 million. Today, the living is simple for the masses, but I had four flat-screen TVs in my Sheraton Towers room in Saigon. The Four Seasons Chiang Mai is very similar to the Four Seasons Hualalai on the big island of Hawaii. You can envision major adjustments to come over time as democracy slowly sways into play and socialism is replaced with free enterprise.

A point to understand is that there is a world of difference between official national policy and the hearts/minds of the people. In both countries there is a mistrust of China and Russia. Amazingly enough, they like America. Our B-52s rained hell over Cambodia to get at the Viet Cong, killing a million of their citizens (note that, in comparison, "only" 58,000 Americans died in the Viet Nam War), and they still like the USA. The people of Thailand have no problem with China, Russia or the U.S. Also, as nice as everyone is in Thailand, Cambodians are over the top friendlier.

Some historians are now suggesting that the embarrassment of being booted out of Vietnam was an important step in helping bring about the demise of the Soviet Union,[312] for they got overextended even in this part of the world. China subsequently began to soften when we left Southeast Asia, to the point where pure commercialism and prosperity could well someday fracture their form of government.

Thus, not only is there considerable economic and political progress in this part of the world. What is happening here will bode well for humanity into the 22nd century.

Comments (0): It surprised me that was nary a comment.

---

311 http://www.sacred-destinations.com/cambodia/angkor-thom

312 http://rt.com/Politics/2009-12-04/cold-war-lessons-forgotten.html

The following article was submitted to the Huffington Post during my India visit. I don't know why, but it was never published. There is nothing controversial here. Boring, maybe.

Is India Incredible?

Absolutely yes, and appropriately so, because there are so many incredible things here, some good, many not so. My previous visit was to Mumbai when it still was Bombay, and the poverty was oppressive. This has not changed. The following is #3 of my around the world odyssey, following South Korea and Southeast Asia.

I encountered a couple of problems finding my way through customs[313] when I arrived, and experienced a few shocks just touring around,[314] but you can't really appreciate what you have until you see how different things can be. Some are contented to live in a secure cocoon, and the older I get, the more I can appreciate that attitude. But, I'm here as tribute to my wife,[315] which began with my HuffPo on "Gratitude, Not Grief."[316]

The people of India have been around for 9000 years. No one is sure when their primary religion, Hindu, started, but, certainly more than 1000 years before Christianity. Muslims conquered the country around 1200, with the grandson of Genghis Khan, Babur, establishing the Mughal (Islamic) Empire in 1526, which lasted until 1857, when the British came. At this point, the country had dropped to 84% Hindu (today, it is closer to 80%, plus 14% Muslims).

After nearly a millennium of occupation by Muslims and Christians, India finally gained independence in 1947, with Pandit Jawaharlal Nehru as their first Prime Minister (PM). The Father of the Nation, Mohandas Karamchand (Mahatma) Gandhi, was assassinated a few months later at the age of 79. During this period, bowing to religious freedom, Pakistan (West and East) was created, inducing 10 million mostly Muslims to relocate. In 1971, East Pakistan became Bangladesh, a third country.

The president, who is voted into office, is the head of state, but the PM, appointed by the ruling parliamentary party, has political control. Presidents have been untouchable (Kocheril Raman Narayanan) and female (Smt. Pratibha Devisingh Patil), now currently in office.

Nehru's daughter, Indira Gandhi (not related to Mahatma), later became PM, but was assassinated. Rajiv Ghandi, eldest son of Indira, became the youngest PM, but also was assassinated. The current PM is Manmohan Singh, the first Sikh to hold this post.

---

313 http://planetearthandhumanity.blogspot.com/2010/02/india-sucks.html

314 http://planetearthandhumanity.blogspot.com/2010/02/india-is-experience.html

315 http://planetearthandhumanity.blogspot.com/2009/08/in-honor-of-pearl.html

316 http://www.huffingtonpost.com/patrick-takahashi/gratitudenot-grief_b_241390.html

Hindu is most widely spoken, but English is remarkably not. I've seen numbers in the range of 20%, but don't really don't know what that means. Eighteen major languages with 1600 dialects are used. Can you imagine what the voting ballots look like? There are supposedly at least 36,000 different newspapers and somewhere between 1800 and 4000 dailies, depending on who you cite.

The country has a little more than one billion people, and is expected to pass China around 2030. Just the increase during this period will amount to about the population of the United States. The number of people is their biggest problem. Their population density is almost three times that of China and ten times the USA.

But the economy is booming,[317] and in percentage growth, could overtake China this year.[318] However, this is where population comes into play, for the GDP/capita of India is $2100, while that of China is $6500, compared to $46,400 for the U.S.

There remains a caste system. You want to be a Brahmin (about 50 million are), but untouchables still number 170 million. Attempts are continuing to end discrimination, with some ministers and even a president (1997, Narayanan) were untouchables. In fact, click on The Present Realities of Brahmins in India,"[319] and clearly, change is occurring.

The cost of living is low. A harrowing 45 minute taxi ride from the airport to my hotel was all of $5, and I paid $50 for a full day (16 hours) tour from Delhi-Agra to the Taj Mahal and Red Fort, with an excellent Indian lunch. There were three of us, accompanied by a guide, driver and navigator. You ask, why navigator? You've heard that time slows down for someone on the verge of a cataclysmic accident? This happened at least a dozen times that day to me, for the navigator signals when a driver can pass, and how he made it formed a series of miracles, some major. I survived the 10 hour roundtrip. This was memorable beyond belief, and I already am a better person for it.[320]

My simple solutions for India?

1. Reduce population. There are too many people. However, I should point out that the world birth average is 3.0 children/woman, and while Afghanistan is 7.9, this figure is 2.8 for India,[321] as compared to 2.05 in the USA, 1.81 China, 1.29 Russia, 1.26 Japan and 1.08 South Korea. Politics/religion will make a one child policy impossible. Nature, in the form of a deadly virus or devastatingly hot summer through global warming, perhaps. But realistically, all India can do is to improve their economy, and wait for the benefits to follow. This will, unfortunately, take a couple of generations, or many more.

---

317 http://www.menafn.com/qn_news_story_s.asp?StoryId=1093286509

318 http://news.bbc.co.uk/2/hi/8273464.stm

319 =" http://www.hindujagruti.org/news/3779.html

320 http://planetearthandhumanity.blogspot.com/2010/02/india-is-experience.html

321 http://www.nationmaster.com/graph/hea_fer_rat_tot_bir_per_wom-rate-total-births-per-woman

2. More honesty in government. A Corruptions Perception Index[322] of 3.4 is better than other countries in this region (Pakistan 2.4, Russia 2.2, Cambodia 2.0, Afghanistan 1.3), but the U.S. is 7.5 and Sweden 9.2. Start with airport customs and I'll be back someday.

Notwithstanding, I certainly am glad I came, for it was inspirational to see the progress and considerable vitality. The economy is surging and trends look promising. India should attain incredibility by 2100 if my two simple solutions can be attained...and the world can avoid economic doomsday from the combined hammer of Peak Oil and Global Heating.

---

322 http://www.transparency.org/policy_research/surveys_indices/cpi/2009

This might have been my most intense day, ever, swinging back and forth between fear and decadence. I was so happy to leave India, but had one more frightening detour. Then the Lufthansa flight was heaven, landing in a whiteout over Munich, ending 24 hours later in Barcelona. This 15February10 posting says it all. Go to my daily blog site for photos.[323]

Just Another Day in My Life[324]

I earlier reported on my around the world odyssey in two HuffPos: The Wonder of South Korea and Progress in Southeast Asia. I continue on in my tribute to my wife Pearl, covered in Gratitude...Not Grief.

I awake at 5 AM to be limousined to the Indira Gandhi International Airport, where Lufthansa checks me into first class, wow. With some trepidation, though, because in the back of my mind is my posting of "India Sucks,"[325] where I criticized customs in Delhi. But I pass through immigration with ease. I was about to have a croissant and capuchino breakfast in the Lufthansa lounge when a uniformed official asked to see my boarding pass (BP). He then orders me to accompany him, for there was a problem with my check-in baggage. Oh no, maybe another extortion scheme?

I imagined all sorts of worse case scenarios. Maybe someone had somehow snuck in a pound of heroin or a bomb into my baggage. We passed through a security gate, where a second uniformed officer, this one with a rifle, accompanies us, and into the bowels of the airport. I've traveled more than 2 million miles on United Airlines alone, perhaps ten around the worlds, and this had never occurred. Then I thought, yikes, that article must have really pissed off someone, and I was headed before an execution squad. Well, Cambodia, maybe, one of my previous stops on this trip, but certainly not India. To make a long story short, after another half hour of sheer agony, they just wanted to check on a cigarette lighter. But the stress was so intense that upon returning to the lounge, I fixed myself a stiff Bloody Mary.

I then went to my gate, but learned that there was a two-hour delay. No problem, at least I made it this far. Finally I boarded, and **the 8-hour flight was about the best I've had in all my years of flying**.

I was offered a Champagne Veuve Devaux with macadamia nuts. According to the write-up, the "freshness of the chardonnay knits perfectly with the fruit and opulence of the pinot noir." In case you did not know, this combination (with Pinot Meunier sometimes) is what goes into champagne, and this one was the winner of the Lufthansa blind tasting.

---

323 http://planetearthandhumanity.blogspot.com/2010/02/from-delhi-to-munich-to-barcelona.html

324 http://www.huffingtonpost.com/patrick-takahashi/just-another-day-in-my-li_b_461714.html

325 http://planetearthandhumanity.blogspot.com/2010/02/india-sucks.html

In the air, my hostess, Emma, passed out Lufthansa pajamas and three large booklets, one each for the two meals and a third of the beverages. I thereby, then, although Pearl would most certainly not have approved, decided to break the Guinness world record for the most esoteric assortment of alcohols imbibed on one flight.

My first ordered drink was a La Guita Sherry, a Manzanilla. If you're counting, I'm now up to three. With the Serrano Ham and Milan Salami I had a Kaseler Nies'chen Riesling (#4), from the top vineyard in the Mosel Valley. Quoting again: "racy acidity, classic minerality and restrained sweetness." Various cheeses then came, followed by an omelette of smoked salmon and creamy spinach, with pineapple stuffed pancakes. This was yet only breakfast!

I ended the first meal with a Niepoort White Port (#5), which had been aged for 10 years. Yes, white port, which actually is golden yellow. I had a Johnny Walker Blue scotch (#6) with the first movie and Warsteiner Premium Verum (#7) with the second.

The next meal started with real caviar and traditional garnishes, including a special Smirnoff vodka (#8). I reflected on the irony of this feast as the plane flew over the Middle East War, but was troubled by this over the top decadence and mild alarm about my carbon footprint. Paul Theroux of Eastern Star fame would have taken a train, in third class. Maybe I'll initiate a new genre: high-end travel with Robin Leach-like flourishes, plus the occasional Indiana Jones adventure.

I skipped most of the rest of the meal, save for a Pomfret in Saffron Sauce with Fettuccine, accompanied with an Ihringer Winklerberg Spatlese trocken (#9), a Pinot Blanc "with subtle aromas of vanilla, pear and citrus fruit." This all ended with a Pfalz Chardonnay Eiswein (#10), balanced with an expresso, topping this all off with a Calvados Pays d'Auge (#11). Actually, much of all the above were mere tastings. I did not truly consume every drop. You can ask Emma, who was a most gracious and accommodating partner in this epicurean experience. Details with photos can be found at "Planet Earth and Humanity."[326]

The flight continues, and the plane landed in a Munich whiteout. I struggled to ease into the Senator Lounge passing through a mob outside. Most departures had been canceled, and the line to re-process was at least 200, if not 300, yards long. This was not single file. It was about 5 people wide and not moving at all. Amazingly enough, as I look at my continuing flight ticket, Lufthansa had the foresight to re-book me on a later Barcelona flight, and this one was to leave in two hours. I check the board, and, yes, 95% cancellations, but DLH 4484 still there.

Unfortunately, the snowstorm worsened, and the plane did not leave until midnight, five hours late. But this was one of the few flights allowed to take-off. I did, though, need to first delicately traverse 20 yards of 6 inch high powder snow wearing standard dress shoes and a light sportcoat through a blizzard from the bus to the plane. But I just saw thousands in panic wondering what they were going to do, so I smugly tolerate this harsh environment.

---

326 http://planetearthandhumanity.blogspot.com/2010/02/from-delhi-to-munich-to-barcelona.html

Two hours later I landed in Barcelona. In fact, my two bags were the first to appear. Again, just a step before freedom, was asked to have my bags x-rayed in a side room. The setting was uncomfortably perfect. I was the only passenger, they saw a computer in my luggage, and they essentially asked the same question as in India. Theirs was, "how many computers do you have in this suitcase?" I said one. "Is that all," says the official? Yes (although I could have added, I have another one in the other bag, but didn't). He kind of shook his head and I thought, okay, how much do I need to pay? Well, he said fine and helped me place my luggage onto the cart and waved goodbye to me. My relief was overwhelming. Spain's Corruption Perception Index[327] is 6.1, compared to #1 New Zealand at 9.4, and the U.S. at 7.5, versus India, with a 3.4. The rest of my stops will be in countries with a CPI higher than the USA.

This is not quite all. It turned out that I had made a brilliant decision the day before, having given the option of being picked up by the Barcelona Le Meridien for 66 Euros (about $100), but I said no because there was a clause that said if the driver had to wait, it would cost 66 Euro/hour. My plane was 8 hours late, so I would have been charged almost $1000. I ended up paying 40 Euros for a cab, but there is a surcharge after midnight, and I added a generous tip. Twenty fours after waking up today, I have survived India and am now in Spain, exhausted, but euphoric.

Comments (4): Again, not much of a response, but I noticed that most of my friends dread seeing their comments in print. Many sent me an e-mail of awe and enjoyment.

---

327 http://www.transparency.org/policy_research/surveys_indices/cpi/2009/cpi_2009_table

I'm pleasantly surprised that the Huffington Post allows me pontificate on so specific an item as the future of Hawaii. On 19February10 I was inspired to suggest an out of the box concept for my home state. We are in a rut and can't deal with the big picture, for we waste all our time on issues like furloughs. We need to merge our efforts on a common cause to change our mentality.

The Sustainable Expo for 2020[328]

Hawaii, the most isolated major populated area on this planet, is that canary in the coal mine of Peak Oil. The economy is so locked into the visitor industry, that the coming jump in oil prices will mean skyrocketing jet fuel prices and the end of tourism as we know it.

You would think that with this so obvious inevitability the State would by now have forged a plan to avoid this calamity? Nope. As pointed out in "We Need to Work Together, Now,"[329] politics, union-labor relations and personality clashes have overwhelmed good sense. Maybe worse, there appears to be no sense of urgency.

Is the problem beyond realistic salvation? Absolutely not, for all we need to do is diversify our economy, accelerate a next generation aircraft (as, perhaps, one powered by hydrogen)[330] and develop a bio jet fuel.[331] There is that worry about timing, of course, because the former has been attempted, and went nowhere, and the latter two will take time, many, many decades, in fact, and Peak Oil could happen tomorrow. Well, to be a bit more optimistic, world oil futures only show crude up to $95/barrel into December 2018, and I personally know respected advocates who feel that both a sustainable aircraft and competitive jet fuel from algae can be developed in time.

Yet, many of my local colleagues have already entered a survival mode, and feel the best they can do is to properly educate the masses about this doomsday scenario. I've been chided for advocating false hopes. But giving up cannot be an option. Fortunately, monumental solutions can best come when there is sufficient desperation. So, what then?

I am today midway on an around the world adventure to seek solutions for our global society, as reported in various **Huffington Post** and **Planet Earth and Humanity** postings. Unexpectedly, this broad search resulted in clarity regarding my home town, for I noticed that a few cities I have visited found a way to transform themselves. The Olympics, in particular, served as this catalyst. In Seoul, Barcelona, Munich and Helsinki, if not for this event, they

---

328 http://www.huffingtonpost.com/patrick-takahashi/the-sustainable-expo-for_b_468009.html

329 http://planetearthandhumanity.blogspot.com/2009/07/we-need-to-work-together-now.html

330 http://www.greenpacks.org/2010/05/17/futuristic-hydrogen-powered-aircraft-specializes-in-vertical-take-off/

331 http://www.guardian.co.uk/environment/2010/feb/13/algae-solve-pentagon-fuel-problem

probably would not have attained their present greatness. World Expos[332] can also work, for from Montreal to Shanghai (begins in May),[333] what happens is that people begin to work together for a common cause. The experience is enlightening, for they learn that rivals in unison can, indeed, accomplish miracles. Cooperation leads to success reinforcing credibility cascading into a progressive municipality.

Okay, Hawaii can forget any Olympics. A World Expo? Hmm ... maybe. This is hardly a new idea, but the conjunction of Peak Oil, Global Warming and a range of impacting factors suggests that the timing is now.

What is needed is clear and imaginative leadership. Usually, some unknown has triggered this all in these resurgent cities. That individual had those qualities to involve power brokers and gain consensus. Then, somehow, the effort gained a life of its own.

Hawaii has a few wild cards in this mission. For one, this is the most beautiful spot in the world. Everywhere I've been, just about everyone thinks there is a certain magic about these islands. Two, we have a host of billionaires, some relatively young, who have already shown a high sense of community activism. Now, if some of them can somehow work together, they should be able to transcend politics, unions and personalities. Three, if we don't do anything magnificent, soon, it will all be over.

Since I've come this far, let me suggest a theme and, further, how to optimize the infrastructure, paving the way towards a Hawaii of the 21st century that can indeed be that worthy dream, as featured in my final chapter of *Simple Solutions for Humanity.* We are in the middle of the largest ocean. The expo can set the stage for the utilization of the riches of the sea for clean energy, green materials and exciting habitats in harmony with the marine environment.

Lockheed Martin is supposedly designing a 100 MW ocean thermal energy conversion[334] facility, and the timing would be ideal to feature this technology. More so, that floating platform could house a major resort, perhaps with a, shudder, casino. A mature Disney at Sea, perhaps. Yes, a Blue Revolution.[335]

A dozen years ago Lisbon hosted what they termed as an Ocean Expo, but there was nothing... nothing, of any nautical moment. Nearly a quarter of century before that, Okinawa showcased Aquapolis.[336] Say 2020 is selected as the target year, the timing would be ideal for another ocean expo.

To link with the "Lockheed Martin/Disney" platform, rather than set aside a plot of land and have various countries build structures that will become obsolete, there already are twenty

---

332 http://en.wikipedia.org/wiki/Expo_(exhibition)

333 http://en.expo2010.cn/a/20081116/000004.htm

334 http://www.huffingtonpost.com/patrick-takahashi/the-coming-of-otec_b_145634.html

335 http://www.huffingtonpost.com/patrick-takahashi/the-dawn-of-the-blue-revo_b_145889.html

336 http://www.tensinet.com/database/viewProject/3843

ideal locations awaiting inspiration. Honolulu is suffering through the pangs of planning a mass transit system.[337] Funding crises will no doubt appear, again and again. Why not find a way to allow international teams to finance, design and manage these stations? Like in EPCOT Center around a lake, each site would feature a different region of the world interfacing, in principle, with the Pacific Ocean. A China village, with the architecture, restaurants and entertainment otions of that country. Maybe they'll bring and leave two pandas. Same for Japan, Korea, the European Community, South America, Africa...and more.

Yes, provide each world partnership development rights and free access to space for a century within a hundred yards of each station, with an ocean corridor, as appropriate. Condominiums and even hotels might be included. Certainly, a few corporate outreach centers, for everyone else will only be a mass transit ride away.

The traditional naysayer, and we really have too many of them these days, will predictably argue that we will be selling Hawaii to the highest bidders, and worse. We need to look on this as our final opportunity to attain supreme world class status. That other option, I remind you, could well be a continuous local depression.

Heck, why not have each island also form international teams and spread the exhibit state-wide. Thus, **Hawaii Sustainable Expo 2020: Visions for Blue Planet Earth**. When the official exposition runs its course, nothing really ends. The State rather suddenly is transformed into the international gathering place for pleasure, blue-green development and global commerce.

Comments (5): Excellent feedback. Expand from Honolulu to the entire state and "think global, act local." Now what?

---

337 http://www.honolulutransit.org/

*On 27February10 I submitted my most memorable posting. That morning there was a gigantic earthquake off Chile that triggered a tsunami. The **Huffington Post** contacted me and asked for an article about the incoming Hawaii tsunami. Terrific I said, although I was in Amsterdam. No problem, as CNN showed two feeds, one from Waikiki Beach and the second of Hilo Bay. I actually could see more about Hawaii than most who lived there because they had to evacuate to the hills. Before the tsunami was scheduled to strike Hilo, I submitted the piece. They immediately published it. However, minute by minute, things change. The value of a virtual newspaper is that you can adjust your report. Thus, over the next hour I changed the following posting at least a dozen times. Remarkably, people who read this report linked it to my blog site, Planet Earth and Humanity.[338] I normally receive a hundred hits a day. That day, 3,356 visited with me.*

Hawaii Tsunami?[339]

I'm on an around the world odyssey, which you can follow through my **HuffPost** postings.[340] Well, my most exciting day is happening right in front of me watching the 8.8 Chile earthquake cataclysm on CNN from the Hotel Pulitzer in Amsterdam. Thus, from half a world away, I'm missing what I've long imagined: a graphic view of a major tsunami decimating Honolulu from my penthouse.

Let me clarify, for that is the last thing I would want. Fortunately enough, the latest scientific data seem to suggest a tsunami perhaps from one foot to seven feet striking Hawaii. But you never know, for the 8.5 Chile earthquake of 1922 caused havoc in Hilo Harbor, and 8.8 is three times more powerful. The underwater topography surrounding Hilo seems to amplify the effect, as history has shown.

The media have reported that the Chile earthquake was 63 times more powerful than the 7.0 Haiti quake, while others have mentioned a thousand times stronger. The reason for this discrepancy can be attributed to a change in the system. The Richter Scale was long the standard, and it turns out that 8.8 is about 60 times the "shake," or horizontal amplitude, of a 7.0 earthquake. However, the current official reporting system is Moment Magnitude, and while the power factor from a 7.0 to 8.0 increases by a factor of 10 using the Richter Scale, you need to multiply by 31.6 for the Moment Magnitude Scale. Thus, for the latter, a 9.0 is 31.6 times 31.6, or 999 (call this a thousand) times the expended energy of a 7.0 earthquake.

Thus, the Chile 8.8 is about 600 times stronger than the Haiti 7.0 using the Moment Magnitude Scale. However, the most powerful earthquake known occurred close to the present one in 1960, a 9.5, which generated a tsunami that killed 61 in Hilo, Hawaii.

---

338 http://planetearthandhumanity.blogspot.com/2010/02/chile-earthquake.html

339 http://www.huffingtonpost.com/patrick-takahashi/hawaii-tsunami_b_479530.html

340 http://www.huffingtonpost.com/patrick-takahashi

According to GMA News,[341] a 6 foot tsunami struck French Polynesia. Reports are coming in from various Pacific sites of similar wave amplitudes.

I'm now watching Hilo, live, with all ships at sea, for if you are sufficiently offshore, the wave will not be perceptible. However, Hilo Bay is now projected (11am Hawaii time) to be struck by a six-foot tsunami. You can also go to *Planet Earth and Humanity*[342] and get the latest info.

Remember, though, that there can be a second and third wave, and sometimes, the later ones are more significant. Well, the Hilo tsunami is just happening, with waters appearing to be receding off Coconut Island at 11:19AM, just exactly as originally predicted. In any case, as a tsunami travels as fast as a jet plane, Honolulu is 30 minutes away (before Noon).

The worst part of this all (at 11:30AM) is the uncertainty. Those sirens beginning at 6AM, and repeating every half an hour or so, are unnerving. The scientific reports seem somewhat reassuring, but there is discoloration in the waters around Hilo and the Wailua River seems to be rising.

Well, now getting close to noon, and the 35-foot tsunami of 1960 will not be repeated in Hilo. At noon, unofficial inputs report on a 2 foot recession and 3 foot rise on the Big Island. So, yes, Hawaii is being impacted by a tsunami, but the all clear signal came. The latest news, though, can still be found at Hawaii Tsunami Information.[343] Twenty two hours after the earthquake, the tsunami also caused some minor harbor problems in Japan.

By the way, to close this discussion, it turns out that the maximum height of a tsunami in the far field (1000 miles away or more) is about 30 feet. However, if there is a landslide of sufficient volume, moving fast enough, into deep waters, the resultant tsunami could be 300 feet or higher. At least, that is a theory that was somewhat exhibited in Lituya Bay (Alaska) in 1958. A wave of 1700 feet![344] My chapter 6 of *SIMPLE SOLUTIONS for Planet Earth* provides a possible worst case scenario for Seattle if such an event occurred where Hilo, of all the ironies, fell into the sea. Details can be found beginning with my 11September08 blog.[345]

Comments (24): There were a lot of comments. I can see where the world wide web can become a useful tool during certain natural disasters. Somehow, I need to translate this instantaneous feedback to remediate Peak Oil and Global Warming.

341 http://www.gmanews.tv/story/184959/tsunami-causes-damage-on-french-polynesia-islands

342 http://www.huffingtonpost.com/patrick-takahashi/hawaii-tsunami_b_479530.html

343 http://hitsunami.info/

344 http://www.doomdaily.com/2009/the-1700-foot-tusnami-that-struck-alaska/

345 http://planetearthandhumanity.blogspot.com/2008/09/six-hours-to-seattle-part-1.html

There seems to be a universal sense that government is broken, and we need something else. Government 2.0[346] is just one of many, and my essay on World 2.0 touched on that subject. Here is a follow-up, published on 3March10.

One Global Government[347]

Six weeks into my global odyssey, I am at Heathrow Airport awaiting my flight back to the USA. Various *HuffPos*[348] have reported on my journey.

Yes, I've had some memorable meals, met fascinating people and continued my tribute to Pearl.[349] Certainly, this trip has reinforced that old bromide: you can best appreciate your life by experiencing how bad it can be. Poverty, corruption, environmental degradation, airline strikes and natural disasters...they all made me stronger and better.

But the one pervading thought was, wherever I went, how badly government was broken. From local municipalities to the U.S. Congress to the United Nations, all signs point toward a declination of our lifestyle because of their inability to govern. Of course, when you come down to it, the real problem, no doubt, is us, the masses.

Remember that Great Recession? Combined with the dual hammer of Global Warming and Peak Oil, the coming economic turmoil can only be worse, much worse. I have already lamented on the "canary in the coalmine"[350] position of my State of Hawaii.

That is the fatal flaw of our society: we just cannot make any necessary decisions until it is too late. The resultant agony from another global depression, attack of aliens from Mars or some truly deadly virus, appears to be necessary before united action can be galvanized.

This all led me to envision our future governing mode. The United States began as a ragtag bunch of colonies, and in only a little more than two centuries, became the most powerful, ever. Re-inventing democracy was brilliant. Of course, India probably dreamt up this concept perhaps 10,000 years ago, remaining as the largest practicing nation today, and Greece refined the form, although it is pathetically not the Athens of the ancient past.

---

346 http://www.manhattan-institute.org/government2.0/

347 http://www.huffingtonpost.com/patrick-takahashi/one-global-government_b_481917.html

348 http://www.huffingtonpost.com/patrick-takahashi

349 http://planetearthandhumanity.blogspot.com/2009/08/in-honor-of-pearl.html

350 http://www.huffingtonpost.com/patrick-takahashi/the-sustainable-expo-for_b_468009.html

The USA will have no serious competition for many generations, China is getting old before it will get rich[351] and Russia is just plain declining,[352] especially in population. Viet Nam will have more people in not many more years. Joel Kotkin, in his "The Next Hundred Million,"[353] provides some reinforcement of this superiority. Thus, the presence of only one superpower can provide some stability as the world attempts to save itself.

The European Union (EU) formed in 1993 with the Maastricht Treaty, seeing how well our united concept succeeded. Mind you, they are today a mess,[354] but, still only a teenager, and are a work in progress. We have almost exactly the same Gross Domestic product and they have 500 million people to our 308 million.

There are some definite strengths, for the Corruption Perception Index[355] lists 7 of the top 11 countries from Europe (U.S. is #19), and, again, 7 of the 10 best world cities in the Mercer Quality of Living Survey[356] are from Europe (Honolulu is America's best at #29).

While some have hinted of this coalition fracturing, for only 16 of the 27 members use the Euro, with resistance growing, chances are that several nations will still be added and even the United Kingdom will eventually join in full. Give them another century.

The next consolidation will be an Asian Alliance (AA), every important country touching the West Pacific, plus India. They will have a population advantage more than five times that of the EU. In fact, Indonesia has more Muslims (200 million) than any other country in the world and more than the core of the Middle East, and China's economy has been predicted to surpass that of the U.S. as early as 2020,[357] or by 2035.[358] Sure, China could instead still go the way of the Soviet Union. Nevertheless, over time, one Asia will prevail, for economic reasons.

Russia is a particular wild card in this crystallization process, for they belong both in the EU and AA. They could well become that key link to meld the two into one someday.

Then, why not a true United States of America, incorporating Central and South America? The Middle East? In a decade or two they will begin to run out of oil and water. While the G8 nations mostly dabble on economics, they have at least considered poverty in Africa and climate change, but they will need to get very serious soon about this ominous Islamic clash

---

351 http://fistfulofeuros.net/afem/economics/getting-old-before-getting-rich/

352 http://www.russiablog.org/2007/06/russias_declining_population_w.php

353 http://www.npr.org/templates/story/story.php?storyId=123497650

354

355 http://www.transparency.org/policy_research/surveys_indices/cpi/2009/cpi_2009_table

356 http://www.citymayors.com/features/quality_survey.html

357 http://www.nyconsulate.prchina.org/eng/xw/t257301.htm

358 http://www.usnews.com/news/articles/2008/07/08/chinas-economy-is-on-track-to-surpass-the-us-by-2035.html

of civilizations. It should be remembered, of course, that there are more than 1.5 billion Muslims in 150 countries.

The trend, though, is clear. First a G3, then G1, a universal government. This already is happening with one global economy. We are inextricably linked on one Planet Earth and our common environment. World peace will only come when there is one global government.

Let's for the sake of discussion say that someday the world has one government. Won't this be like the United Nations, but worse? Well, linear thinking would come to that conclusion. Let us surprise ourselves and be smarter this next time. Let us design a system that works.

Yet, it remains unclear what will trigger this ultimate consolidation. Certainly, the United Nations (UN) will not morph into this leading role. As the EU has 27 members, should the U.S. then have 50, not one vote? If you've ever had to work with this organization, you can only agree with me that the UN has perfected paralysis by analysis. Thus, the UN cannot be that unifying force.

So what is the simple solution on how, exactly, to attain one world government and Peace on Earth? I really don't know. Perhaps someone out there has a clue.

In any case, the whole point of seeking total cooperation through, maybe, one government, is to end all major wars forever. A glimmer of hope is a pathway I suggested as a strategy for now President Barack Obama in my very first Huffington Post article,[359] when he was still competing for the Democratic nomination nearly two years ago. Click on that posting and let me know if you have a better idea.

Comments (10): The comments were varied, but mostly supportive. I made contact with an individual who especially like the 10% solution suggested in my first HuffPo, and we are communicating.

---

359 http://www.huffingtonpost.com/patrick-takahashi/well-barack-we-have-a-pro_b_104201.html

This was a shock. Another rather boring posting on government, but there were 85 comments. I have yet to figure out why. Anyway, I wrote this 11March10 article late one night, probably inspired by my e-mail exchanges with James Dator, resident futurist at the University of Hawaii.

Evolution, Devolution, and Revolution[360]

We all know what evolution is, except that only 40% of Americans believe[361] in the Darwinian kind. Not particularly surprising, but only 21% of those who did not complete high school believe in evolution, while 74% of those who have at least some kind of postgraduate schooling do.

Devolution in the federal sense is the return of rights to states, but for the purpose of this discussion, let us use the biological definition, that is, backward evolution, or for society, a lowering of our lifestyle. Before the Grand Recession, I would guess that less than half thought that, perhaps, our civilization had peaked and will now decline. Today, I would not be surprised if more than half of Americans have an uncomfortable feeling that the combination of Peak Oil, Global Warming and our broken government is so severe that, while our economy will soon get better, there is a distinct possibility that we might have experienced the best, and future generations, beginning with our children, will only see a decline in their future life.

I continue to be one of those who feel that giving up is not an option. Sure, after the second energy crisis in 1979 we should have known better and thirty years ago initiated that crusade to convert from the fossil fuels to renewable energy. Keep in mind, though, that the Greenhouse Effect was only in the vocabulary of academics, even though Svante Arrhenius[362] in 1896 already began warning of global warming. He was the first Swede to win a Nobel Prize, and as President Barack Obama was in the latest group, you must know that this award is given in Sweden. Anyway, a primary reason why we never got our act together three decades ago is that in 1998, the real price of oil actually dropped below what it was in 1973[363] when the First Energy Crisis shocked us. In fact, even lower than compared to 1946.

Thus, don't unnecessarily blame oil companies, the White House and Congress for what you allowed, with good reason. This all comes back to who is responsible and all my **HuffPos** have blamed us, the masses. That famous Pogo quote,[364] "We have met the energy, and he is us," has a universal application, and first appeared almost 60 years ago.

---

360 http://www.huffingtonpost.com/patrick-takahashi/evolution-devolution-and_b_495007.html

361 http://www.gallup.com/poll/114544/darwin-birthday-believe-evolution.aspx

362 http://www.lycos.com/info/svante-arrhenius.html

363 http://www.inflationdata.com/inflation/inflation_Rate/Historical_Oil_Prices_Table.asp

364 http://www.igopogo.com/we_have_met.htm

Okay, so, what then? Let us do something about what we created. We elect our political leaders, buy gasoline and electricity, and should have total control of our future. Something wonderful has happened to allow us to orchestrate this preferred future: the world-wide web. Media like **The Huffington Post** can now make that crucial difference. We need to initiate a new revolution to take back control of our future.

So what do we do? I don't know! If you believe in scientific revolution (and unfortunately, Americans are ahead of the curve, for way far less than half of our world population does), evolution got us this far about 250,000 years ago. Humanity blossomed, with almost 7 billion people today. We sidestepped the potential nuclear winter about two decades ago when the Cold War ended, only to run headlong into the last few years of cheap oil, a climate change potential that largely precludes the use of coal, the evils of current fission and a governing system unable to act with dispatch.

Well, on the more positive front, there is fusion, the process our Sun and all the stars utilize to create energy. What about the Hydrogen Economy? Or, maybe, even signals from outer space, yes, the Search for Extraterrestrial Intelligence, my chapter 4 of ***Simple Solutions for Humanity.*** I'm stretching, but I'm getting desperate. So should you.

We can either quickly educate the general populace about the coming devolution, or initiate this revolution. Let us start by, indeed, letting the people know about those bad case scenarios, for what are the odds of actually galvanizing a whole new way of thinking to save Planet Earth and Humanity? Mind you, we don't really need to convince 7 billion souls. We have already proven to be sheep. We must get to those key revolutionary influencers, a handful of individuals and organizations, I think. Don't be patient, for time is a dimension that is crucial to our fate.

But what exactly do we do? Again, I don't really know. My most recent and 70th *HuffPo* on "One Global Government," provides a clue, and my first *HuffPo* on "Well, Barack, We Have a Problem," at least suggests a pathway. I am an almost insignificant and certainly lonely voice in the ether, and can only be a catalyst. I hereby pass on this world saving charge to you, although, I pledge to assist in any way I can.

Comments (85): To quote:

The purpose of this posting was to spark a virtual revolution towards finding real solutions for the double hammer of Peak Oil and Global Warming, as our various governments seem broken and can't seem to make any necessary decisions. We instead largely veered into various energy discussions, okay, but not quite the galvanization I sought. But I learning something and gained a better sense about how to next proceed.

Alas, no galvanizing solution to save Planet Earth and Humanity.

This 14April10 posting was a follow-up to a couple of my virtual forums. Someone came up with the idea that we needed something like the Union Concerned Scientists' Doomsday Clock, not on nuclear energy, but for Planet Earth and Humanity, incorporating, the economy, peak oil, global warming and all the other parameters influencing life as we know it. Somehow, our group, or a representative panel, or just input from the people, would provide a number (or something) that captures the imagination of the general public. The adjustment would be continuous as the input arrives into the computer. Well, the following is a first thought at some points to consider.

The Index of Life[365]

Our species, scientist term *Homo sapiens*, has been around for around 100,000 years. How long is 100,00 years? Well, light would take 100,000 years[366] just to cross our Milky Way galaxy. Our closest neighbor, the Andromeda Galaxy, is 2.5 million light years away from us. The distance of one light year is just under 6 trillion miles. While these numbers merely represent the vastness of our corner of the Universe, it is humbling to consider these space and time dimensions as we worry about today.

Anyway, we've survived to dominate, but struggled through some tough times. About 75,000 years ago, for example, the world population dropped to only a few thousand according to the Toba Catastrophe Theory.[367] Another threat to our civilization occurred a little more than 6 centuries ago when up to half the European population were killed[368] by bubonic plague. But our population only dropped,[369] at most, from 450 million down to 350 million. Not a big deal in terms of extinction.

The greatest threat to humanity, no doubt, was only half a century ago, when the Cuban Missile Crisis in 1962 edged our society to the brink of an extinctive nuclear winter.[370] We overcame our greatest test, for the end of the Cold War in 1991 might well have insured for our continuation in perpetuity. I'm taking the super optimist's viewpoint that the Sun will not change much for a few billion years and fusion or whatever will someday enable our kind to populate the Universe.

Technological progress has been remarkable, and there is a sense that our lifestyle will only get better and better. A terrorist dirty bomb will be messy, but our continuation as a species will not be threatened by this act, nor any attempt by Iran and North Korea to assert themselves.

---

365 http://www.huffingtonpost.com/patrick-takahashi/the-index-of-life_b_536619.html

366 http://www.economicexpert.com/a/Light:year.htm

367 http://www.sciencedaily.com/articles/t/toba_catastrophe_theory.htm

368 http://www.learner.org/interactives/renaissance/middleages.html

369 http://en.wikipedia.org/wiki/Black_Death

370 http://www.treehugger.com/files/2010/01/nuclear-winter-easier-to-trigger-than-previously-thought-study.php

But here is the problem. More and more I'm beginning to be influenced by some of my doomsday colleagues who have become frighteningly convincing that the combination of Peak Oil and Global Warming means that life will, from now, only get worse. Your children, and, certainly, theirs, they say, will descend into mediocrity, and worse. Renewable energy? Not sufficient, according to their analysis. Toss in hopeless governance, and they might have a few good points.

It is entirely possible that humanity on Planet Earth might have attained the peak of life in the year 2000, a nice and symbolically strategic round number, being the turn of the millennium, but also, a year before the presidency of George W. Bush and 9/11/2001. As an economic reference point, the value of the Dow Jones Industrials at this pivotal time in history was similar to what it is today.

A book by Richard Heinberg entitled *Peak Everything* provides greater detail. In short, it is conceivable that the **Index of Life** (IOL), on "average," will only decline, forever. Let me arbitrarily give the year 2000 a 10 rating, and drop the IOL to 5 for where we were from October 2008 to March 2009 (our Great Recession). In short, a range of parameters is involved in this Index, but, frankly, the survival and enjoyment of life pretty well represent the essence of that number. How, then, is Humanity doing today? Here is a summary:

1. **The Economy**: While most fiscal pundits remain careful, I say that we are out of our Great Recession. There will be no serious double-dip, I hope. While the National Bureau of Economics last week announced it was not still sure, the UK officially declared this clearance status in January and our Department of Commerce hinted so in October last year. Why such optimism? The Dow Jones this week ended above 11,000 for the first time in 18 months; the Federal Budget Deficit for March 2010 dropped by a factor of three from a year ago; the unemployment rate, which was supposed to be well above 10% through much of this year, seems to want to remain at a single digit; our Gross Domestic Product is predicted to grow between 3% and 5% this year; IPO's are back and doing well; housing can improve, but shows signs of recovering; General Motors is progressing and the less than $2/share Ford I purchased a little more than a year ago shot past $13/share today; and banks are doing well, with loans now generally available. Remain concerned about the European PIGS,[371] but from the brink of depression, it looks now as if the Democrats might not lose too many House and Senate seats this Fall.

2. **War and Peace**: Five years ago, the United Nations unanimously voted to outlaw nuclear terrorism.[372] While this is about as close to a useless vote as there can be, at least the tone was set to what has transpired this week. President Obama welcomed 47 countries to his DC Nuclear Security Summit, and in parallel, UN Secretary General Ban Ki-moon urged a ban on producing fissile material for weapons, indicating that the 65 nation Conference on Disarmament in Geneva will develop the plan. Then, too, Iran announced that China will

---

371 http://my.opera.com/richardinbellingham/blog/portugal-italy-greece-and-spain-the-pigs-of-europe-are-causing-an-ouflow-of

372 http://www.washingtonpost.com/wp-dyn/articles/A51708-2005Apr13.html

be among the 60 countries to meet in Tehran next week to discuss nuclear disarmament. A little one-upsmanship here? All this on the heels of the amazing momentum catalyzed by a President Obama and President Dimitriy Medvedev Nuclear Arms Reduction Pact signed on April Fool's Day, and unfortunately so. Thus, worry a bit about terrorists, Israel eviscerating an Iranian nuclear site within two years, and North Korea going haywire in that same timeframe...but the end of the Cold War wiped out the danger of Mankind destroying ourselves with nuclear bombs. Now, if President Obama can only adopt my plan proposed in my original *Huffington Post* article.

3. **Peak Oil**: I just read a blog article entitled, "After Peak Oil, Are We Heading Towards Economic Collapse?"[373] Remembering that the $147/barrel oil in July of 2008 could well have triggered this Great Recession, and the early '80's price of oil up to an equivalent today of $97.50/barrel (or $191/barrel relative to GDP share) was clearly responsible for the 1982 world economic collapse, we nevertheless tend to lose sight of the fact that petroleum has caused those economic traumas this past half century. Yet, the Global Futures price of oil[374] in December 2018 has been dropping for the past few months, and is now at $96/barrel. So, this is good news, for, perhaps, the world is being given time to try to recover. But how often have oil price prognosticators been right?

4. **Global Climate Change**: The past decade has been the warmest on record. We keep hearing this again and again, year after year. On the other hand, disinformation specialists have been euphoric with the coldest winter in recent times for Siberia. In February, I traipsed through Finland and much of Europe, where everyone kept telling me of the worst snowfalls in twenty years. But how can the hottest and coldest occur within the same year? Is global warming an elaborate hoax for scientists and Al Gore to gain continued funding and Nobel Prizes? No, and even most skeptics will concur that some element of warming is occurring. It is just that no one can convince decision makers that we are causing it. Thus, the naysayers say why prematurely spend trillions, and those concerned say if we don't act now, it will only cost more in the future. We today have a dangerous stalemate, and nothing much will be done until tens of millions perish one hot summer. Then, of course, there is that fear of even worse: destabilization of the dynamic equilibrium keeping methane at the sea bottom, sparking THE VENUS SYNDROME.[375]

So, all in all, our economy is improving; world peace is nowhere close to reality, but the world at large is safe from cataclysmic termination; peak oil seems to have been pushed off into the future; and global warming policy will remain in limbo. Thus the Index of Life for today I place at 8.0. Comments?

Comments (10): A few responses, but no virtual revolution.

373

374 http://www.cmegroup.com/trading/energy/crude-oil/light-sweet-crude.html

375 http://www.huffingtonpost.com/patrick-takahashi/the-venus-syndrome-part-o_b_106120.html

My 20April10 posting seemed like a good idea at the time. However, not long thereafter, that Icelandic volcano stopped being a pest.

Volcanoes: Hawaii Versus Iceland[376]

On January 3, 1983 I was golfing at the Volcano Golf and Country Club on the Big Island of Hawaii, when, on our back nine, the ground shook and we saw fountains of lava. It was a Kilauea Volcano eruption that continues until today, 27 years later. I cite this experience because Eyjafjallajokull (E),[377] that Icelandic volcano which erupted on March 20, continued on for 14 months when last active in 1821. Worse still, every time this volcano has gone off (only three in a millennium), its much larger sister, Katla (K),[378] has also erupted. The combination of **E** and **K** could well be truly ominous. While air travel is beginning to return, and this current phase could end today, what will happen to Europe if this episode continues for years?

Iceland, like the islands of Hawaii, was formed by volcanoes. In 1783, Laki (L)[379] killed half the livestock and a quarter of the population, and in 934 Edlgja[380] might well have had the largest basalt flow in the history of Planet Earth. Oh, there are 35 active volcanoes on Iceland.

AccessScience[381] has an excellent summary of volcano eruptions and the potential impact on humanity. ***Simple Solutions for Planet Earth*** describes various potential natural disasters, including the mega hyped Cumbre Vieja volcanic eruption said to be capable of creating a mega tsunami. Further described is a similar scenario where a portion of the Big Island of Hawaii falls into the ocean, with the potential for generating a mega (normal amplitude max is 40 feet, while mega starts at 40 meters and could go up to 500 meters and higher) tsunami.

Located on the Big Island, Kilauea is the most active and visited volcano in the world. It is so accessible and "safe" that the Hawaii Volcano Observatory actually began doing science right on that mountain nearly a century ago. It's still there.

When the wind blows from the Big Island towards Honolulu, we become Los Angeles of 1960 when the smog was truly terrible. I have canceled golf outings when these days occur.

---

376 http://www.huffingtonpost.com/patrick-takahashi/volcanoes-hawaii-versus-i_b_543571.html

377 http://globaleconomicanalysis.blogspot.com/2010/04/iceland-volcano-eruption-forces.html

378 http://www.huffingtonpost.com/dk-matai/risk-of-katla-could-a-2nd_b_541755.html

379 http://www.kwintessential.co.uk/articles/article/Iceland/Laki-Volcano-Eruption-Iceland/529

380 http://en.wikipedia.org/wiki/Eldgjá

381 http://www.accessscience.com/abstract.aspx?id=735300&referURL=http%3a%2f%2fwww.accessscience.com%2fcontent.aspx%3fsearchStr%3d%22Volcanic%2bash%2band%2baviation%2bsafety%22%26id%3d735300

There is no question in my mind that lifelong exposure to the Hawaiian vog must damage lungs, much more so on the Big Island than other islands. The concrete wall structures on my roof are turning a shade of yellow (from the sulfuric acid fallout) and small black particulates are visible, which can only mean that your lungs breathe in these micro lava shards.

Okay, the situation in Iceland is different, but here in Hawaii, we have taken advantage of this potential calamity: Kilauea eruptions increase the visitor count! A few surveys[382] have been undertaken, and the general data shows that cardiorespiratory influences can be detected, and crops have been damaged. This is not a particularly big deal in the media, partially, I worry, because the economy of Hawaii is fully dependent on tourism, and as poorly as the industry is currently doing, a condemning conclusion could be yet another dagger that can sink us into a local depression.[383]

While as earlier mentioned, our eruptions, with rare exceptions, tend to mostly flow, with sulfurous gases, Icelandic ones, perhaps because of the overlying glaciers, are much more explosive, tossing a lot more particulates into the air. As a result, the bankruptcy in Iceland might well turn out to be contagious if **E** is joined by **K** (no, not Kilauea, but Katla, and sure, why not add Laki, plus Edlgja), and continues belching. Thus, the European Union can add to the PIGS list yet another economic thorn. Worse, if prolonged, the health of citizens could be affected. So learn how to pronounce EYJAFJALLAJOKULL[384], for like in Hawaii, this natural disaster could well hang around for some time to come.

Comments (4): Not much of a response.

---

382 http://www.ncbi.nlm.nih.gov/pubmed/19092552

383 http://www.huffingtonpost.com/patrick-takahashi/the-sustainable-expo-for_b_468009.html

384 http://abcnews.go.com/GMA/video/pronounce-eyjafjallajokull-10392613

My 17June10 posting is explained in the comments below.

Human Cloning: A Post-Life Insurance?[385]

My *HuffPosting* on "Gratitude, Not Grief"[386] almost a year ago led me to consider a variety of funeral options. I learned that cremation will soon overtake burial. As a futuristic thinker, a Swiftian thought occurred to me that a logical next step might be preservation of a small portion of the remains just in case cloning someday becomes commonplace. A post-insurance policy?

While there are already cryogenic processes,[387] there is no assurance that this technique can have any happy ending, especially with the added costs involved. Then again, success was attained with frozen embryo births.[388] Forty-three years ago James Bedford was placed in deep cold storage[389] and baseball player Ted Williams[390] was cryogenically preserved in 2002.

On a mass scale for the public at large, though, it should ultimately not take much to keep a cubic centimeter of bone marrow, drop of blood or piece of your skin. Certainly, some technique can be devised at low maintenance cost to preserve your DNA/RNA specimen.

Remember Michael Crichton's *Jurassic Park*? Recall that the blood of a dinosaur, sucked by a mosquito, which happened to get enveloped in amber, kept for a 50 million years at room temperature, became the viable fluid biologically cloned into a prehistoric beast. Well, that was an exaggeration of reality, but Japanese and Russian scientists are now in the process of cloning a woolly mammoth[391] using this technique. Thus, it is not a stretch of wild imagination for a small sample of your dead body to be carefully stored, for, who knows, it could just be a matter of a few decades before science might perfect this process for humans.

There will be fly-by-night outfits to take your money, but the potential of a credible business succeeding around the world will happen, and probably soon. What organization already into searching for roots has the reliability, managerial ability, international roots and entrepreneurial capability to capture the marketplace? This "company" should have long term credibility, having endured controversy, but nevertheless remain thriving. The Mormon Church came to mind. Already, Ancestry.com[392] is taking cheek-swiped cotton swab and

---

385 http://www.huffingtonpost.com/patrick-takahashi/human-cloning-a-post-life_b_613661.html

386 http://www.huffingtonpost.com/patrick-takahashi/gratitudenot-grief_b_241390.html

387 http://www.cryonics.org/cryostats.html

388 http://en.wikipedia.org/wiki/Cryopreservation

389 http://www.alcor.org/Library/html/BedfordSuspension.html

390 http://www.cryonics.org/ted.html

391 http://blogs.discovery.com/animal_news/2009/05/cloning-the-mammoth.html

392 http://www.ancestry.com/

$200 to add DNA to your family tree. Yes, Sorenson Genomics,[393] the firm doing the work, is headquartered in Utah.

I can visualize repositories resembling a high-tech mini crystal palace. The setting will be a comfortable purgatory where the visitor can punch in a code to electronically view a video of the person at rest.

As the average funeral[394] in the U.S. today costs $6,500, the price tag of sustaining each person for this new enterprise should be a fraction of that sum. A simple charge of $1000 should be a financial and recruiting windfall for the Church. Analyzing the long-term total benefits, a seeming lost leader charge of *free* might even be considered.

Each year, about one percent of the world population dies. Thus, the potential exists to add 60 million new specimens each year. Every society honors the dead in some respectful manner. If the charge is to be free to secure your loved one, with the additional prospect of perhaps a resurrection someday, that practice could well become universal.

Several billion bodies are already in the ground, and another billion or so stored in urns. While most samples are perhaps not now ideal for cloning purposes, the potential remains, opening up another marketing pathway.

Now, all this might at first blush sound much too sacrilegious and sardonic, and even offensive to some. If so, delete the religious references. What about then some business organization with the reputation and means to carry out this ghoulish imperative for a profit? This simple solution would make available land for more productive uses, reduce the cost of death and make it more convenient to visit those who have passed on, as many will merely carry out this privilege at home through the internet. And don't forget the ultimate: the possibility of a real afterlife.

Comments (13): *The purpose of this posting is best captured by one of my responses:*

Actually, Ron, I'm not doing any pursuing. What happened was that I submitted an article updating the field of SETI, and the editors of this fine publication chose not to use it, which I found somewhat irritating, for this was good science. I think there remains that certain giggle factor associated with the field. So, I thought I'd test the limits of what is acceptable, and my daily blog happened that day to be serializing a portion of my religion chapter from SIMPLE SOLUTIONS for Humanity. So I submitted "The Future of Graveyards." They changed it to the title above, which, I've got to admit, is more provocative. Sufficiently appeased, I am now completing Part 2 of Human Cloning and should send that in soon.

---

393 http://www.sorensongenomics.com/

394 http://www.ehow.com/about_5132574_typical-funeral-costs.html

I thought the timing was right on1July2010 to crystallize the essence of all my energy and the environment HuffPostings into an article. One simple solution for Peak Oil and Global Warming, which was a 5 cents/pound carbon dioxide investment...or tax.

A Simple Solution for Peak Oil and Global Warming[395]

I have now penned two **Simple Solution** books, one on **Planet Earth** and the second on **Humanity**. Let me draw from the first one and provide just one simple solution to solve our energy/environment problem. But first, some background.

The current issue of **_Time_ (July 5, 2010)**[396] reports on Bill Gates (Microsoft) and Jeffrey Immelt (GE) just last week, representing a group of corporate titans, beseeching Congress to *triple* U.S. spending on energy research. They underscored that energy gets less than $5 billion/year, but $80 billion goes for military R&D.[397] Said Immelt:

"This is about innovation. This is about competition. This is about energy security."

Never have industrial leaders collectively made this kind of plea about sustainable resources over defense. If we can spend $3,000 billion (also known as $3 trillion)[398] on just the Iraq War, supposedly to neutralize Saddam's weapons of mass destruction, but, more to protect oil and bring peace to the Middle East, hindsight argues that if this sum had been applied to spur our private sector to replace fossil fuels with renewable energy, we might not today be at the precipice awaiting the dual hammer of Peak Oil and Global Warming.

My second **Huffington Post** article of two years ago blamed the lack of will on part of the people[399] for our current predicament. This aloof attitude remains, for our masses are now accustomed to $3/gallon gasoline (remember, Europe is double to triple this price), plus, they can't appreciate hardly detectable global temperature increase and sea level rise. It hurts that the disinformation campaign from oil/coal interests and their academic supporters, who a recent survey found to be less credible scientists,[400] are easier to believe than the rantings of a bunch of scientists accused of a possible hoax. But what do you expect from a nation attracted to vampires[401] and the afterlife?[402]

---

395 http://www.huffingtonpost.com/patrick-takahashi/a-simple-solution-for-pea_b_631774.html

396 http://www.time.com/time/specials/packages/0,28757,1999143,00.html

397 http://www.huffingtonpost.com/patrick-takahashi/why-do-we-spend-so-much-o_b_116535.html

398 http://www.laprogressive.com/the-middle-east/the-3-trillion-war/

399 http://www.huffingtonpost.com/patrick-takahashi/why-is-there-no-national_b_104507.html

400 http://environmentalresearchweb.org/cws/article/news/43003

401 http://goldderby.latimes.com/awards_goldderby/2010/06/twilight-eclipse-twilight-saga-374281695-news.html

402 http://www.huffingtonpost.com/patrick-takahashi/evolution-global-warming_b_168827.html

Swine flu[403] and oil spills bring out the personal concern and general ire on the part of our populace, but energy and carbon dioxide policies are somewhere between ho hum and who cares. The fact that many Democrats in the Senate from fossil fuel states are preventing the Obama Administration from pushing along the Waxman-Markey Clean Energy Bill,[404] which was approved by the House more than a year ago, is just another symptom of how much our decision-makers care about any national energy policy, something we have never had.

As I have underscored in my *HuffPost* entries, I actually don't blame the Republicans, George W. Bush, oil companies or OPEC. They were maximizing their interests, and in a free enterprise system, that's fine. Well, I did have a gripe about President Ronald Reagan when he came into office in 1981, for he decimated the solar budget. However, the real reason why our sustainable resources were never commercialized was the price of oil, an essentially unpredictable commodity. Very few can actually remember that petroleum, in terms of 1998 dollars, was the cheapest, EVER, that year.[405] Go to that article and just trace the red line until you hit the bottom, and see that you're in 1998. Yes, less expensive than just before the First Energy Crisis in 1973. This fickleness will continue to bedevil renewable energy investments.

What responsible financial institution could afford to take the risk of loaning a hundred million dollars for a solar energy project during those days? While the Chicago Mercantile Exchange[406] today predicts a light crude oil future price of $91.65/barrel in December of 2018, would you stake the destiny of our country on that investment figure? Remember, it was only two years ago this month that oil peaked at $147/barrel. But who knows where the current $70-$80/barrel range will go. $150/barrel if Israel bombs Iran? $35/barrel if there is a more serious double dip grand recession?

We thus need to take extraordinary action, and Gates/Immel's proclamation was an excellent start. Here they are, industry barons, actually asking government to spend more money on something not directly related to their profit margin (well, GE does sell wind turbines). The marketplace cannot determine the fate of our world, for it takes a full generation (25 years), and longer, to shift energy sources. *We need to start today.* So, here is my simple solution:

1. **Immediately** enact a 5 cents/pound carbon dioxide credit.[407] Okay, this is the same as a tax, but read my HuffPost article[408] on this subject. This credit will only increase gasoline by a buck a gallon and double the price of coal electricity. Congress needs to pass this measure, which President Obama should sign, and he further needs to have the G8

---

403 http://www.huffingtonpost.com/patrick-takahashi/a-pandemic-worse-than-the_b_207226.html

404 http://energycommerce.house.gov/index.php?option=com_content&view=article&id=1697:house-passes-historic-waxman-markey-clean-energy-bill&catid=155:statements&Itemid=55

405 http://planetearthandhumanity.blogspot.com/2010/06/future-of-energy-part-2.html

406 http://www.cmegroup.com/trading/energy/crude-oil/light-sweet-crude.html

407 http://www.huffingtonpost.com/patrick-takahashi/the-carbon-dioxide-credit_b_221933.html

408 http://www.huffingtonpost.com/patrick-takahashi/the-carbon-dioxide-credit_b_221933.html

nations, China and India comply. Ah, but you say, easy to suggest, but impossible to do. Sure. Simple solutions can be difficult to accomplish. So, what next?

2. Ask Mother Nature to raise the temperature of the atmosphere so high this summer that tens, if not hundreds, of millions perish. On my contention that we have a fatal flaw in our society -- that we cannot make important decisions until it is too late -- we then need a cataclysmic event, the more horrible the more effective. Yes, this is terrible, but, save for those who actually die, this would be like taking some bitter medicine to cure your ailment. Then you say, but we have no influence over Mother Nature.

3. Not true! This is all a matter of time. A kind of doomsday will occur if we largely continue on our current consumption pathway, for more species will become extinct, weather will go haywire, and humanity will interminably suffer before action is finally taken. Clearly, our decision-makers will not have the courage to just do it, and the American people just do not riot in the streets for this sort of cause. I was kind of hoping that this world wide web mechanism would more directly replace marching protests, but, I haven't yet figured out how to catalyze response.

4. So what then? Await *The Venus Syndrome*,[409] or at least the upcoming novel of that title. Three degrees F rise? Try an increase of 800 degrees and the end of life on Planet Earth. Waiting might not be a viable option.

<u>Comments (66)</u>: *In many ways, this posting was a tossing of all my energy and environment HuffPostings into a summary. Why it drew so many responses is a mystery to me, but it could have been my provocative retorts, such as:*

We generally agree on most points, including dual purpose military R&D. However, many of us do play God and control our immediate space. The problem comes when there is conflict with others. I remain hopeful that humanity will eventually mature into a true partnership for the benefit of most. We might not be 7 billion then, but Planet Earth should be fine and sustainable for, say, a billion people. Then, in a millennium, there is outer space to consider. The Sun will not overwhelm our globe for another 5 billion years or so, giving us time to be fruitful and multiply.

That might have been as doomsday as I get. Mind you, as is the usual case, many of the comments were supportive, and I sometimes reply with something like:

Hey, thanks. You did, in particular, hit the sweet spot by mentioning revolution. Somehow, I think, virtual portals such as the *Huffington Post* should be able to replace those oh so last generation protest marches. Any thoughts on how to galvanize and catalyze using HuffPo?

---

409 http://www.huffingtonpost.com/patrick-takahashi/the-venus-syndrome-part-o_b_106120.html

On 6July2010 I felt compelled to re-visit the notion of world peace, forever. This was my third try at waking people up to this issue, of paramount interest to U.S. Senator Spark Matsunaga. As you will see, I'm not doing this right.

The 10% Simple Solution to Peace[410]

What are the truly monumental problems facing our society and how do we fix them? I dealt with one, Global Climate Change, in my previous *HuffPo* post.[411] Just a 5 cents/pound carbon dioxide credit (or tax) is all we need to save Planet Earth and Humanity. At this writing, there were 64 comments, and while most were supportive, a few just did not get the point. I was using sarcasm and fear instead of almost useless pure logic.

On this Independence Day when I am clicking this article, I thought it was particularly appropriate to take a closer look at the second most important issue: war and peace. When I worked for U.S. Senator Spark Matsunaga, considerable effort was extended towards legislation for the U.S. Peace Academy. Sparky, who earned a Purple Heart in the European Theater, actually still had a piece of shrapnel in his leg. He felt that if the Nation had all those war universities, why not a U.S. Peace Academy to train emissaries for goodwill. The best he could get was the U.S. Institute of Peace.[412]

We lament the inability for Democrats to work with Republicans in our Congress, but national defense is a strange political beast. Clearly, Republicans, supported by conservatives and the military industrial complex, favor spending more money on defense. Democrats like to help the common folk, and in the $60 billion war funding bill,[413] tacked on $20 billion of domestic goodies last week that will cause considerable heartburn in the Senate. A filibuster, even, has been threatened by Republicans.

This same bill fought back an amendment seeking withdrawal of American troops from Afghanistan. President Obama threatened to veto the bill if this measure was included.[414] Stay with me here. President Obama and General Petraeus agreed that American troops, conditionally, will begin transferring responsibilities to their Afghan counterparts in a year.[415] The House amendment which put those words into a deadline was pushed by liberal Democrats, and beaten back by Republicans, with the aid of Obama. Are you as confused as I am? There is a nuance here that defies normal politics or common sense

---

410 http://www.huffingtonpost.com/patrick-takahashi/the-10-simple-solution-to_b_635285.html

411 http://www.huffingtonpost.com/patrick-takahashi/a-simple-solution-for-pea_b_631774.html

412 http://www.usip.org/

413 http://www.capitolhillblue.com/node/28641/comment-page-1

414 http://thehill.com/homenews/administration/106889-veto-looms-over-war-supplemental-bill

415 http://www.cnn.com/2010/POLITICS/07/04/afghanistan.withdrawal.deadline/?hpt=T1

To some degree, this is all linked to the budget and the world economy. Do we spend more money now to insure against that double dip, or do we curtail spending to reduce the national debt? Nobel Laureate Paul Krugman[416] advocates borrowing and spending more to avoid a third depression (in case you missed it, the first one was the Panic of 1873[417] and the second that Great Depression of 1929). President Obama is of the Krugman school and was somewhat rebuffed when the G8 Muskoka Declaration[418] last month chose to slash spending.

Very few still remember that when President Ronald Reagan assumed office in 1981, the Second Energy Crisis placed him in almost an exact economic position as Obama, and the Congress gave Reagan a $750 billion stimulus package.[419] In 2010 dollars, this equates to nearly $2 trillion. Thus, Krugman's additional trillion dollar stimulus plan[420] to the original $850 billion spending measure only reflects a history that worked.

However, the world and American people have more recently taken a thrifty turn. As more than half our Federal budget goes to the military,[421] and we have no obvious enemy (there are only a few thousand terrorists worth worrying over) into the long term future, why not cut our defense outlays? This might be the ideal moment in time to lay the foundation for a more peaceful world of tomorrow.

I have said this before, in my original **HuffPosting**, "Well, Barack, We Have a Problem...,"[422] and "The Ten Percent Solution."[423] To summarize, President Obama goes to the next G8 summit, and pronounces a Gorbachev-like bombshell: "America will reduce defense spending by 10% this coming year, and will continue to slice 10% every year if you all do the same." In just a very few years, military spending will be minimal and the world will be at a higher level of peace forever.

This is the 10% simple solution to peace. Before you make any inane comments, click on those articles to appreciate that Russia is getting feeble and China will also become old before it gets rich. The U.S. will be supreme for a long time to come, and those war funds can better be applied to cure Planet Earth and enhance the fate of Humanity.

Comments (5): The subject of PEACE just has no reading interest. This might well be the most important long term goal of humanity, but no one cares. There were only five comments, three

---

416 http://pajamasmedia.com/richardfernandez/2010/06/30/crystal-balls/

417 http://www.thehistorybox.com/ny_city/panics/panics_article9a.htm

418 http://www.voltairenet.org/article166076.html

419 http://www.maysvillekybbs.com/forums/showthread.php?t=10299

420 http://newsbusters.org/blogs/jeff-poor/2010/06/28/krugman-tries-scare-more-government-spending-third-depression-rhetoric

421 http://www.warresisters.org/pages/piechart.htm

422 http://www.huffingtonpost.com/patrick-takahashi/well-barack-we-have-a-pro_b_104201.html

423 http://www.huffingtonpost.com/patrick-takahashi/the-10-solution_b_168090.html

mine and two from friends who were mostly nice to respond. I am having a minimal effect trying to sell renewable energy and warn about the dangers of global warming, which is a lot more productive than my peace effort.

On 7July2010 the Huffington Post allowed me to focus on Hawaii as the symbol of energy independence.

Hawaii: The Proposed Symbol of Energy Independence[424]

Hawaii is that proverbial canary in the coal mine regarding Peak Oil and the economy. Our only hope is a global partnership to as quickly as possible help us attain a high level of energy independence. But why should Hawaii be singled out for this privilege?

The reasons are many, but the most compelling is that we are the ideal sustainability test tube: progressive leaders, abundance of renewable options, high cost of energy (an electricity bill 250% higher than the national average, so commercialization can more quickly be attained), relatively small size (less than one half of one percent the population of the Nation, so the investment will be affordable), singular political clout (the most powerful congressional member in Senator Daniel Inouye, and leader of the Free World, President Barack Obama, who was born in this state) and, soon, sheer desperation, and, therefore, motivation. Provided is a golden opportunity for the World to work together with us to create a symbol for sustainability.

Hawaii is blessed with the Sun, tradewinds, heat of the Earth, ocean resources and sufficient rain to grow biomass on irrigated lands recently vacated by the sugar industry. We pay the highest energy prices in the Nation.[425] Ninety percent of the energy we use is petroleum, which is destined to skyrocket in price when Peak Oil occurs.

How are we doing on the development of renewable energy relative to other states? First of all, EIA 2007 data[426] (2008 information will be released later this month) for the country shows hydroelectric power at 2.4%, wind and geothermal are both at 0.33% and solar at .04%. Wind, particularly, should be higher in 2010, but, still, non-hydro renewable energy remains at the noise level.

I selected a cross section of states[427] (click on Planet Earth and Humanity[428] for the full data) and found that when you deduct hydro, Hawaii is doing as well as California and Colorado, but better than all the others. Yet, we remain 94% fossil fueled.

---

424 http://www.huffingtonpost.com/patrick-takahashi/hawaii-the-proposed-symbo_b_637414.html

425 http://apps1.eere.energy.gov/states/energy_summary_print.cfm?state=HI

426 http://www.eia.doe.gov/cneaf/solar.renewables/page/rea_data/table1_1.html

427 http://www.eia.doe.gov/cneaf/solar.renewables/page/state_profiles/r_profiles_sum.html

428 http://planetearthandhumanity.blogspot.com/2010/07/future-of-energy-part-5.html

For the future, Governor Linda Lingle proclaimed the Hawaii Clean Energy Initiative,[429] touting 70% renewable energy by 2030. This certainly confused me, for also stipulated was 40% of all electricity by 2030 when aviation fuel (about 30% of energy used) might well then still be close to zero and ground transport can only be a huge guess. Conservation is part of this strategy, as it should, but I doubt if we'll use much less energy in two decades.

Hawaii will elect a new governor this Fall, so, in advance, here are my recommendations for his (yes, all three are male) consideration:

1. Your deep sea electric cable project[430] should include the Big Island because geothermal energy, a baseload power source, should be included in the future mix. A quarter century ago I assisted in the planning for a similar venture as there was potential for 500 MW of geothermal.[431]

2. Be sure to determine potential ocean thermal energy conversion[432] (which is baseload) sites so the cable can be conveniently tapped at these locations.

3. This deep sea electrical cable will then cost from $2 billion to $3 billion. Can we afford this? Well, capital improvement projects[433] get close to $2 billion annually, so spread over a decade, maybe. There, too, are bonds to float. However, the most sensible proposition is to have our congressional delegation (plus that of Texas and California) and the White House introduce the National Grid Act of 2011, using Hawaii as the first total system site, with Texas as the wind power demo and California for utility-scale solar. It is now becoming obvious that major wind/solar farms are being delayed because of the cost of an accessible smart grid network. We installed the national highway system in the '50's, and where would we be today without our interstate freeways? Now, a national grid to wheel electricity is clearly our next universal need.

4. A larger problem will be ground transportation. Plug-in electric cars seem to be in vogue today, but a direct methanol fuel cell[434] powered transport system makes more sense in the long run. Immediately delete ethanol and terrestrial biofuels (too slow growing, very inefficient and uses too much water). There, though, is something about microalgae,[435] and research should be significantly expanded on this option.

5. Our most vulnerable lifeline is aviation. Tourism is about our only real industry, and when the price of petroleum jumps to $150/barrel, jet fuel will become so expensive that our

---

429 http://www.renewableenergyworld.com/rea/news/article/2008/11/hawaii-takes-bold-renewable-energy-initiatives-54036

430 http://www.greentechmedia.com/articles/read/hawaii-weighs-undersea-cable-to-deliver-wind-power/

431 http://www.punageothermalventure.com/About-Geothermal-Energy/13/geothermal-in-hawaii

432 http://www.huffingtonpost.com/patrick-takahashi/the-coming-of-otec_b_145634.html

433 http://sunshinereview.org/index.php/Hawaii_state_budget

434 http://www.huffingtonpost.com/patrick-takahashi/is-there-an-option-more-p_b_150824.html

435 http://www.guardian.co.uk/environment/2010/feb/13/algae-solve-pentagon-fuel-problem

tourism rate will drop by 50%. Our local economy will go into and stay in deep depression for many decades, for there are no substitutes on the commercial horizon. Do everything in your power to insure for both a substitute jet fuel[436] and a next generation sustainable aircraft.[437] The former will take more than a decade, and the latter, maybe a generation, and more. We might have time, though, as the Chicago Mercantile Exchange[438] lists the price of crude oil at less than $90/barrel in December of 2018. But how often have we been surprised by sudden spikes? Very few saw $147/barrel oil coming until it happened. And, Peak Oil could well be just around the corner.

I can warn you that TIME should be your greatest concern. If Peak Oil never occurs, Hawaii and the rest of the World will only be thankful. So we guessed wrong. Yet, this embarrassment will mostly result in more locally produced clean energy, which will only help our economy in the future. What are the odds, though, for petroleum to remain under $100/barrel for the next quarter century?

Comments (17): Not much of a response, but satisfying. When I write an article about renewable energy, there is almost a sure chance that the discussion will find several threads, nuclear power and photovoltaics especially. There are dreamers in every resource technology, and either they refuse to accept reality, or are right or more probably somewhere in between.

---

436 http://www.huffingtonpost.com/patrick-takahashi/biofuels-from-microalgae_b_347093.html

437 http://www.huffingtonpost.com/patrick-takahashi/the-future-of-sustainable_b_270969.html

438 http://www.cmegroup.com/trading/energy/crude-oil/light-sweet-crude.html

If you will recall, I wrote my initial cloning posting on 17June2010 because HuffPo chose not to publish my submission on the Search for Extraterrestrial Intelligence. So I wondered how absurd I could write something and still have them publish it. Thus, the title was "Human Cloning: A Post Life Insurance?" Originally I had it as "The Future of Graveyards," but they made the adjustment. Well it was published. So I felt compelled to let the other shoe drop, so I turned in the following posting on 15July2010 as "Human Cloning: Part 2." Again, they changed the title (and they only very rarely edit anything I submit) to what you see below, and the comments were, for me, incredible. At last count, it was up to 224, double my previous high.

Science and the Future of Cloning: Is Immortality Possible?[439]

An MSNBC poll shows that 81 percent of Americans don't believe in the afterlife.[440] Yet, a Pew Forum poll shows that 82 percent do believe in an afterlife.[441] How can two respectable organizations be so different in their surveys? On-line poll crashing, perhaps. Well, 90 percent or so of Americans claim to believe in a God, so chances are the Pew version is closer to reality.

Whether you believe or not, most of us have thought about death, and for many "something" after our present life seems better than a dark eternal gloom forever. Hoping the Bible, Koran and virtually every religious publication are right, let us nevertheless speculate on the biological option, for there is a finite chance that they might all be wrong. I certainly haven't seen anything close to compelling proof.

What is "eternal life?" In one sense, all living creatures today are essentially already immortal. We should be able to, someday, trace ourselves back through 50 billion DNA copyings over 4 billion years to determine our LUCA (lowest universal common ancestor).[442] Our DNA has, thus, had everlasting life. While our species almost became extinct in that Great Toba Supervolcano Eruption of 73,000 BC, where Homo sapiens dropped to perhaps a thousand breeding pairs, we have recovered well, survived the potential nuclear winter of the Cold War, and have no obvious doomsday event on the horizon, except, maybe, for **The Venus Syndrome**.[443]

Of course, we will also live through our children and their children. Plus, the products of our life, such as letters, books, digital photos and statues, will be around long after we expire.

---

439 http://www.huffingtonpost.com/patrick-takahashi/science-and-the-future-of_b_643838.html

440 http://www.secularnewsdaily.com/2010/01/20/msnbc-poll-81-do-not-believe-in-afterlife/

441 http://thegreatrealization.wordpress.com/2008/03/25/82-of-americans-believe-in-afterlife/

442 http://www.actionbioscience.org/newfrontiers/jeffares_poole.html

443 http://www.huffingtonpost.com/patrick-takahashi/the-venus-syndrome-part-o_b_106120.html

However, Woody Allen[444] has expressed a sense that he was not satisfied with immortality through his works, for he wanted to live forever by not dying. Conscious eternal life, if not rejuvenation and reversal, then, is an ultimate goal on the level of world peace and universal happiness. Sounds a bit like Heaven.

There are at least two pathways to continue your presence. One does not involve human cloning. Without going into telomeres and ribonucleoproteins, let me just say that science is actually close to finding and checking the aging gene. Someday, you might be able to take a pill and stop growing old. The question is, can this technique be perfected before you get too old? You can still, then, of course, get killed in an auto accident or through an illness, but that so-called 130 year old lady[445] from Georgia (of the former Soviet Union) could someday be commonplace.

The other is cloning, and there are two kinds: therapeutic and reproductive. The former is almost okay, while the latter is verboten, except in certain countries where the laws are fuzzy. You can expect some future breakthrough in countries where religion is not dominant.

Animal reproductive cloning is old news. Scotland produced Dolly in 1997, with mice in Hawaii (1998), Prometea in Italy (horse, 2003), Little Nicky in the USA (a cat, 2004) and Snuppy in South Korea (dog, 2006). Thus, the concept of reproductive cloning has been proven to be real.

So let's get to reproductive human cloning,[446] laden with legal and moral land mines. I sat in on a seminar by Nobel Laureate Joshua Lederberg almost half a century ago while a student at Stanford, where this concept came up during the discussion. The field has both come a long way, and not really that much, over this period.

The UN General Assembly in August of 2005 did adopt a declaration prohibiting all forms of human cloning. The vote was 87 in support, 34 in opposition and 70 abstaining or absent. But the edict was non-binding. The European Convention on Human Rights and Biomedicine prohibits human cloning, but has not been ratified by most countries. There is, further, a Charter of Fundamental Rights of the European Union, which bans reproductive human cloning, but it has no legal standing.

So where is the USA on human cloning? Human cloning is legal in the U.S.,[447] but there are some Federal prohibitions against research. The George W. Bush regime was especially difficult, and Barack Obama ended the ban on embryonic stem cell research, while remaining opposed to human cloning.[448]

---

444 http://www.brainyquote.com/quotes/quotes/w/woodyallen161239.html

445 http://www1.voanews.com/english/news/lifestyle/Georgia-Claims-Worlds-Oldest-Person-130-years-98026964.html

446 http://www.cbhd.org/content/human-cloning

447

448 http://www.cbsnews.com/stories/2009/03/09/politics/100days/domesticissues/main4853385.shtml

Stanford[449] formed a stem cell institute in 2003 and Harvard[450] initiated efforts to clone human embryos in 2006. They initially were attempting to fund this work with private donors without any government assistance. Mind you, they are not cloning humans, as Harvard would like to harvest stem cells to fight leukemia and diabetes. The University of California at San Francisco announced a similar pursuit. Advanced Cell Technology of Massachusetts is commercializing human embryonic stem cell cloning services.

Some countries have observed the American reluctance to support human cloning research and have taken definite steps. There was South Korea and their scandal.[451] The situation is somewhat foggy in the United Kingdom,[452] as the University of Newscastle in 2005 claimed to clone the first human embryo.[453]

Singapore, a former British colony of 4.5 million people, has entered the competition. For all intents and purposes, while a democracy, it is about as close to a benevolent dictatorship as there exists today. The government decides what is best and gets the job done. Biotechnology is a priority area. They created Biopolis,[454] a $300 million, 2 million square foot research center focused on biomedical development, recruiting world class scientists, some who were fed up with the national politics in their own country. Singapore is trying to establish a world sanctuary for stem cell research. While first inaugurated in 2003, Biopolis is already home to scientists from 50 nations. While reproductive human cloning is banned, I can see this island someday becoming the site of choice for therapeutic cloning, as depicted in a former CBS television drama *Century City*.[455]

What about China? Is China[456] a cloning paradise? University of Connecticut animal cloning director Jerry Yang Xiangzhong[457] told *The Standard*, China's business newspaper, that China can jump ahead of the U.S. in three years if their scientists were given the green light to proceed. His contention is that in much of the developed world scientific progress in this field is hindered by political and religious debates. There is also the moral problem with something called human dignity. Apparently, these difficulties would not be experienced in China. Tragically, Professor Yang passed away last year at the age of 49.

Okay, let's say someday human reproductive cloning is attained. The concern always comes up about what good this is, as I won't know this will be the real me. Well, it has been speculated that by the time all these bioethical hurdles are cleared, computer technology will

---

449 http://www.nature.com/nm/journal/v9/n2/full/nm0203-156b.html

450 http://www.msnbc.msn.com/id/13167778/

451 http://news.bbc.co.uk/2/hi/asia-pacific/4554422.stm

452 http://www.bioethics.ac.uk/index.php?do=topic&sid=5

453 http://news.bbc.co.uk/2/hi/health/4563607.stm

454 http://protomag.com/assets/biopolis-the-science-factory?page=5

455 http://bioethics.net/journal/j_articles.php?aid=664

456 http://www.asianews.eu/news-en/China,-a-cloning's-paradise-2653.html

457 http://www.thestandard.com.hk/news_detail.asp?pp_cat=11&art_id=39588&sid=12537776&con_type=1

be developed to the stage where your memory can be transferred to this new body. The field now exceeds 100 trillion calculations per second (1014 cps), and should be at least ten times faster in a decade, at which capability the brain can be simulated.[458] Such a computer should only cost about $1000 in 2020.

That's not all. There are algorithms and biological interfacing challenges. This field is just beginning, but the odds are, this fantasy for immortality could be possible in 25 years.

Finally, the cost factor. Originally only billionaires might be able to afford eternal life. So if you were worried about exacerbating our already overpopulated world, economics, as they are already affecting birth rates, will also check the growth of human reproductive cloning. However, while we all know how Moore's Law has precipitously dropped the price of computing power, the reduction of genome sequencing costs has been a lot more dramatic,[459] so immortality could well be closer than you think.

Part 1 of Human Cloning[460] appeared last month. Chapter 2 of Simple Solutions For Humanity (see icon below) covers eternal life. This series has to do with the afterlife. Chapter 5 of my book goes into a nationally popular topic in afterlife discussions -- the religious afterlife.

Finally, click on
http://www.huffingtonpost.com/david-stipp/why-anti-aging-science-re_b_644308.html
*for a **HuffPo** on anti-aging science. David Stipp recently wrote a book **entitled THE YOUTH PILL**.*

Comments (224): The response was overwhelming. Most were nice and complimentary, and I remarked in response to one:

Thank you for your thoughts and compliments. I sometimes wonder why I even bother to post anything, exposing myself now and then to angry diatribes. However, you have given me yet another reason to continue.

It occurred to me, however, that I might have shown an attitude that ticked off some readers. So I actually apologized:

It has been almost a week since this posting, so the comment period might be soon to end. Again, I thank those that responded. I did learn a few things and even adjusted my article to accommodate new and better information.

I also must apologize for, now and then, I was a bit too sarcastic or humorous or whatever. I suspect this attitude had more to do than the controversial nature of the topic to galvanize a

---

458 http://en.wikipedia.org/wiki/Predictions_made_by_Ray_Kurzweil

459 http://www.technologyreview.com/biomedicine/24590/

460 http://www.huffingtonpost.com/patrick-takahashi/human-cloning-a-post-life_b_613661.html

few angry comments. In addition, there was so much input that it was difficult to keep up, and, in retrospect, I probably would have responded differently given more time and thought.

There must be a clue somewhere in all this to affect productive change regarding Peak Oil, Global Warming and peace. Most of my ***HuffPo*** responses are more supportive than anything else, but lack in passion and commitment. I still think there is something to these real time portals that drives me to improve the process.

Anyway, mahalo and aloha.

*Rather than go on and on, I urge you, if you are interested, to find this **HuffPo** and read those comments.*

This leads me to 2August2010 when this manuscript became due to Author House. Fortuitously, this final one is all about Pearl. This was my one-year remembrance, and a fitting conclusion to this book, as it is dedicated to her.

Life After Death[461]

Coincidentally following up on my *HuffPo* on cloning and the afterlife,[462] this entry is an entirely different view on the concept of life and death. It has been a year now since my wife passed away, captured in my *Huffington Post* article entitled, "Gratitude, Not Grief."[463] That six-week period was gut-wrenching, where she was essentially in an induced coma and there was absolutely nothing I could do.

I remember breaking away one day to see the Academy Award winner, *Hurt Locker*. Suffice to say that my mental state was such that I would have eagerly traded roles with the bomb defusing team in Iraq over my present agony in this paradise called Honolulu.

The mental anxiety was overwhelming, but there were also a few unexpected physical effects. For the past few years I had made an honest attempt to shed six pounds and couldn't. Without realizing it, I weighed myself one day toward the end of this ordeal and found I had lost 11 pounds.

After 47 years of marriage, where, among other things, she looked about the same as when we first met and was admired as the nicest person anyone knew -- plus the fact that we were best friends -- her sudden departure should have been devastating. Interestingly and amazingly enough, I experienced just the opposite reaction.

I wouldn't quite say I have been in constant euphoria, but, surprisingly enough, something very close, for if she had survived, she probably would only have been kept alive through a ventilator. This would have been terrible for her and for the family and me. Her death was a parting gift, for she gave me a new life.

When I wake up every morning, my first scene is Kauakini Hospital in the background, where all of this occurred. Just my imagining that I no longer needed to trudge over to be by her side in intensive care provides the spark that gets me going. My attitude is further enhanced by a combination of a new free will and total independence. I only have a few years left and can do anything I want, including roasting a turkey,[464] another *HuffPost*.

---

461 http://www.huffingtonpost.com/patrick-takahashi/life-from-death_b_664539.html

462 http://www.huffingtonpost.com/patrick-takahashi/science-and-the-future-of_b_643838.html

463 http://www.huffingtonpost.com/patrick-takahashi/gratitudenot-grief_b_241390.html

464 http://www.huffingtonpost.com/patrick-takahashi/how-to-roast-a-turkey_b_366778.html

There was no healing process as such, but any sense of guilt--that I continue living, while she didn't, as our fondest wish was to somehow die together holding hands--is being eased by a purposeful mission to establish The Pearl Foundation[465] as a tribute to her.

Pearl had a special love of a specific yellow tree,[466] which was complicated because there are at least half a dozen such varieties, and, blame it on global warming, they sometimes now bloom more than once a year. I finally decided to search for the scientific name in the Orient, for the one that seemed to stand out was the golden tree, the *Cassia fistula*, the national tree of Thailand. Strangely enough, this national tree does not grow in Bangkok. I met with Hunsa Punnapayak, Head of Research at Chulalongkorn University, who indicated that he sees what appeared to be something similar to Pearl's tree on his way to Pattaya. He indicated that he would seek input from forestry colleagues he knows to identify the exact species.

I then went on to Delhi because I had a second mission. Every morning when Pearl awoke growing up in Hilo, she could see Mauna Kea,[467] Hawaii's tallest volcano at 13,803 feet. She thus exhibited paintings and photos of this mountain in our apartment. Thus, part of her ashes were spread on this mountain.[468] I decided that I would continue the ceremony at a few sites around the world where she had wanted to visit, but never did. Thus, I went to the Taj Mahal after Thailand, where I picked a perfect spot and had a private ceremony.[469] My incredible India stopover was death defying.

Finally, finally, Pearl's sister-in-law, Gwen Nakamichi, found the scientific name and the tree on the Big Island of Hawaii. Called *Cybistax donnelli smithii* (CDS)[470] I purchased two at a Big Island nursery and have them now planted on my roof.

My plans are to give one sapling and my book, **SIMPLE SOLUTION Essays**, dedicated to Pearl (cover the same color as that yellow tree), which will be published by September, to everyone who donated to the Pearl Foundation. I have also initiated discussions with the Arbor Day Foundation[471] to have them include the *CDS* in their free tree giveaway each year. In addition, there is a chance that agreement can be reached with the City and County of Honolulu to plant a bank of these trees on the mountain side of the Ala Wai Canal. Finally, there is a possibility that a Boy Scout troop will take on a project to plant them at a site on the Big Island.

On the second mission, my fall around the world journey will include one additional ash spreading ceremony. I will perhaps post an article on this event after I leave that country.

---

465 http://planetearthandhumanity.blogspot.com/2009/09/pearl-foundation.html

466 http://planetearthandhumanity.blogspot.com/2009/10/more-on-pearls-yellow-tree.html

467 http://hvo.wr.usgs.gov/volcanoes/maunakea/

468 http://planetearthandhumanity.blogspot.com/2009/08/in-honor-of-pearl.html

469 http://planetearthandhumanity.blogspot.com/2010/02/india-is-experience.html

470 http://planetearthandhumanity.blogspot.com/2009/10/pearls-yellow-tree-is-identified.html

471 http://www.arborday.org/

There have thus been eight such sites, which should increase to about a dozen by next spring, after which I plan to write a book on my experience.

So from death can come a new life. I feel re-energized as never before and look forward to the coming adventures.

Comments (16): Thank you, Pearl.

Epilogue

Will there be a Book 4? Yes. With a co-author or two, I will next delve into fantasy and publish a novel on *The Venus Syndrome*.[472] The plot will focus on 21December2012, the date of that overblown Mayan prophecy. However, in a country that loves to watch vampire films and believes in an afterlife, maybe a few will actually enjoy reading our version of doomsday…or not.

---

[472] http://www.authorhouse.com/BookStore/ItemDetail.aspx?bookid=46634